Polymer Conformation
and Configuration

POLYTECHNIC PRESS
OF THE
POLYTECHNIC INSTITUTE OF BROOKLYN, BROOKLYN, N. Y.

CURRENT CHEMICAL CONCEPTS

A Series of Monographs

Edited by LOUIS MEITES

H. J. EMELÉUS: The Chemistry of Fluorine and Its Compounds.

FRANK A. BOVEY: Polymer Conformation and Configuration

In Preparation

P. ZUMAN: The Elucidation of Organic Electrode Processes

HENRY TAUBE: Electron-Transfer Reactions of Complex Ions in Solution

A Polytechnic Press of the
Polytechnic Institute of Brooklyn Book

Polymer Conformation and Configuration

Frank A. Bovey

Bell Telephone Laboratories, Inc.
Murray Hill, N. J.

ACADEMIC PRESS   NEW YORK   LONDON   1969

ACADEMIC PRESS, INC.
111 Fifth Avenue, New York, New York 10003

United Kingdom Edition published by
ACADEMIC PRESS, INC. (LONDON) LTD.
Berkeley Square House, London W.1

LIBRARY OF CONGRESS CATALOG CARD NUMBER: 71-84946

PRINTED IN THE UNITED STATES OF AMERICA

FOREWORD

This is one of a series of monographs made possible by a Center of Excellence grant from the National Science Foundation to the Polytechnic Institute of Brooklyn in 1965. That grant enabled the Institute's Department of Chemistry to establish a Distinguished Visiting Lectureship that is held successively by a number of eminent chemists, each of whom has played a leading part in the development of some important area of chemical research. During his term of residence at the Institute, each Lecturer gives a series of public lectures on a topic of his choice.

These monographs arose from a desire to preserve the substance of these lectures and to share them with interested chemists everywhere. They are intended to be more leisurely, more speculative, and more personal than reviews that might have been published in other ways. Each of them sets forth an outstanding chemist's own views on the past, the present, and the possible future of his field. By showing how the facts of yesterday have given rise to today's concepts, deductions, hopes, fears, and guesses, they should serve as guides to the research and thinking of tomorrow.

This volume is based on a series of four lectures given by Dr. Bovey while in residence at the Institute in January and February, 1967. It is with great pride and pleasure that we present this record of the stimulation and profit that our Department obtained from his visit.

LOUIS MEITES, *Editor*
*Professor of Analytical
Chemistry*

F. MARSHALL BERINGER
*Head, Department
of Chemistry*

Author's Preface

This book arose from a series of four lectures delivered at the Polytechnic Institute of Brooklyn, where the author served as Visiting Professor in January and February, 1967. At certain points, the discussion has been brought somewhat more up to date by the addition of new material, but is essentially as presented at that time.

The emphasis is experimental and the treatment deals primarily with the investigative methods which I and my colleagues at the Bell Telephone Laboratories have employed: high resolution nuclear magnetic resonance (NMR) spectroscopy, optical rotatory dispersion (ORD), and circular dichroism (CD). The discussion has the general theme indicated by the title, but consists of two somewhat disparate parts. The first three chapters deal with the stereochemistry and conformation of vinyl polymers, and with the application of NMR spectroscopy to their study. The fourth and fifth chapters are devoted to studies of polypeptide conformation by NMR and optical methods, including the familiar helix-coil transition in α-helical polypeptides and the somewhat less well known *cis-trans* transitions in N-substituted polypeptides.

The discussion does not pretend to be a complete or balanced one, but deals to a large extent (though not exclusively) with the author's own researches. No apology is offered for this imbalance, as it is consistent with the original purposes of the lecture series.

F. A. BOVEY

Contents

CHAPTER I

The Configuration of Vinyl Polymer Chains

1. *Molecular Symmetry and its Observation by NMR.* In this chapter, we shall examine the fundamental basis of the study of the stereochemistry of vinyl polymer chains by NMR. In order to do this, it is convenient to begin with a consideration of small molecules that are related in structure to vinyl polymers.

NMR is uniquely powerful in providing information concerning the symmetry of molecules of uncertain structure or configuration. Such information often permits a definite decision among possible structures, and may also define bond angles and conformational preferences. The observation of molecular symmetry—or lack of symmetry—depends upon the fact that otherwise similar nuclei which occupy geometrically non-equivalent sites in a molecule will, in general, be magnetically non-equivalent* as well, and will exhibit different chemical shifts and different couplings to neighboring nuclei. Environmental differences too subtle to detect by other means are often obvious in the NMR spectrum. It may,

---

*We do not necessarily use the term "magnetically non-equivalent" here in the sense often employed by the NMR spectroscopist, *i.e.,* to refer to the nuclei of any group which are equivalent among themselves geometrically, and therefore have the same chemical shift, but are unequally coupled to the nuclei of another group. Thus, in *cis* or *trans* HFC=CHF, the protons and fluorine nuclei each constitute a group of non-equivalent nuclei, whereas in CH_2F_2 the members of each group are magnetically equivalent. In our discussion, some nuclei will be called equivalent which would be termed non-equivalent in this more rigorous usage.

of course, happen that nuclei which are geometrically non-equivalent have fortuitously equal chemical shifts or couplings (or both), within experimental error. In this paper, we shall use the term "non-equivalent" to denote the fundamental geometrical fact, regardless of fortuitous experimental equivalence.

Important symmetry problems from the viewpoint of polymer structure are presented by molecules of the type

$$
\begin{array}{c}
\text{M} \\
| \\
\text{X}\!-\!\text{C}\!-\!\text{Y}^* \\
| \\
\text{M}
\end{array}
$$

where X is a group which has a plane or axis of symmetry, such as phenyl, methyl, halogen, etc., and Y* is a group having in itself no element of symmetry. M is any observable atom or group, such as H, F, phenyl, methyl, CF_3, etc. In such a molecule, the groups M will be non-equivalent. The molecule is not necessarily asymmetric as a whole, however, for Y* may, for example, be

$$
\begin{array}{c}
\text{P} \quad \text{M} \\
| \quad\ | \\
-\text{C}\!-\!\text{C}\!-\!\text{X} \\
| \quad\ | \\
\text{Q} \quad \text{M}
\end{array}
$$

in which case the molecule has a plane of symmetry:

$$
\begin{array}{c}
\text{M} \quad \text{P} \quad \text{M} \\
| \quad\ | \quad\ | \\
\text{X}\!-\!\text{C}\!-\!\text{C}\!-\!\text{C}\!-\!\text{X} \\
| \quad\ | \quad\ | \\
\text{M} \quad \text{Q} \quad \text{M}
\end{array}
$$

Such a molecule nevertheless satisfies the requirements for steric differentiation of the geminal M groups in each pair. To simplify the ensuing discussion, we shall refer to all such CM_2 groups in which the M groups are equivalent as *homo-*

steric, and to those in which the M groups are non-equivalent as *heterosteric*.†

The terms asymmetric and dissymmetric have acquired special meanings with reference to optical rotation. A molecule is asymmetric if it possesses no element of symmetry, and is dissymmetric if it possess a symmetry element (most commonly a two-fold axis) but is not superimposable on its mirror image. We have just seen that a molecule which is geminally heterosteric need not be asymmetric. Conversely, an asymmetric molecule may be homosteric. In addition, a geminally heterosteric molecule need not be dissymmetric, and *vice versa*. Within a restricted class of molecules, relationships among these three symmetry types can be found, but there is no entirely general relationship between geminal homostericity or heterostericity on the one hand and asymmetry or dissymmetry on the other.

Some specific examples may make these points clearer:

(a) Y^* is an asymmetric carbon atom, $C-R_1R_2R_3$

For each optical isomer (not ordinarily distinguishable by NMR), there will be three staggered conformers (Fig. I-1),

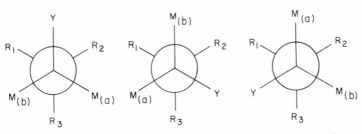

Fig. I-1.   Staggered conformers of substituted ethane $Y-CM_2-CR_1R_2R_3$.

†The author is indebted to Prof. Murray Goodman for the suggestion of this terminology (*cf.* "Topics in Stereochemistry," Vol. II, p. 73, ed. by N. L. Allinger and E. L. Eliel, John Wiley and Sons, New York, 1967). Mislow and Raban ("Topics in Stereochemistry," Vol. I, p. 1) have offered the terms "enantiotopic" and "diastereotopic" to describe homosteric and heterosteric groups, respectively.

and in each the environments of M_A and M_B are non-equivalent. Most of the observed differentiation between M_A and M_B probably arises from an energetic preference for one of the conformers[1] but it can be readily shown that in principle some degree of non-equivalence must persist even if all three are equally populated.[2] A great many examples of this type are now known.

(b) Y* is (see above):

$$-\underset{\underset{Q}{|}}{\overset{\overset{P}{|}}{C}}-\underset{\underset{M}{|}}{\overset{\overset{M}{|}}{C}}-X$$

Many examples of this type are likewise known, *e.g.,* ethyl groups of sulfites,[2-6] sulfoxides,[2,7] diethylmethylammonium ion,[7] acetaldehyde diethyl acetal,[4] and sulfinic esters,[8] and

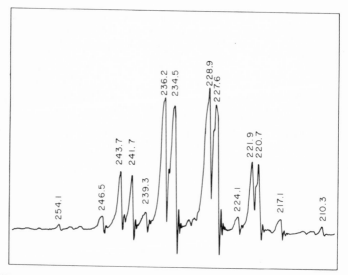

Fig. I-2. Methylene proton spectrum of diethyl sulfite; 50% solution in benzene, 60 MHz [from F. Kaplan and J. D. Roberts, *J. Am. Chem. Soc.,* **83,** 4666 (1961)]. Peak positions are expressed in cps from TMS.

benzyl groups of 3,3-dibenzylphthalide,[9] to mention only a few. The study of heterosteric methylene groups has almost become a separate field of organic chemistry in itself. (We might note that the central atom is not necessarily carbon as indicated.)

Figure I-2 shows the CH_2 spectrum of diethyl sulfite.[6] It is complex because of a chemical shift difference of approximately 0.10 p.p.m., combined with both geminal and vicinal coupling. If the molecule were geminally homosteric, as is diethyl sulphate, the CH_2 resonance would be simply a binomial quartet, but the asymmetric nature of the sulfite group discriminates between the protons in each methylene group.

Figure I-3 shows the CH_2 spectrum of 3,3-dibenzylphthalide.[9] It is an AB quartet because the two protons are in nonequivalent environments, differing in chemical shift by 0.11 p.p.m. and with a geminal coupling of -14 cps.

(c) $X = Y^* = CR_1R_2R_3$, i.e., *the molecule has similar asymmetric centers:*

$$
\begin{array}{ccc}
R_1 & M & R_1 \\
| & | & | \\
R_2 \!-\! C^* \!-\! C \!-\! C^* \!-\! R_2 \\
| & | & | \\
R_3 & M & R_3
\end{array}
$$

If X and Y* are of the same handedness we have either the *d* or *l* diastereoisomer; the racemic mixture of *d* and *l* will usually be indistinguishable from either alone by NMR, and this diastereoisomer is often loosely referred to as *racemic*, whether resolved into enantiomers or not. The molecule is dissymmetric, since it has a two-fold axis, but the CM_2 group is homosteric. If X and Y are of opposite handedness, we have the *meso* diastereoisomer, which of course has a plane of symmetry but in which the CM_2 group is heterosteric. Molecules of this type are clearly very closely related to vinyl polymers. The most carefully studied examples are *racemic* and *meso* 2,4-disubstituted pentanes. Tiers and Bovey[10] showed that the fluorine nuclei of the central CF_2 group of *meso*-$CF_2Cl \cdot CFCl \cdot CF_2 \cdot CFCl \cdot CF_2Cl$ have differing chemical

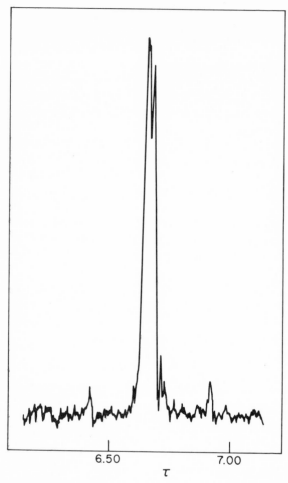

Fig. I-3. Benzyl proton spectrum of 3,3-dibenzylphthalide, 60 MHz [from G. C. Brumlik, R. K. Baumgarten, and A. I. Kosak, *Nature,* **201,** 388 (1964)].

shifts and couplings, while those of the *racemic* isomer are identical. Doskocilova,[11] Doskocilova and Schneider,[12] Satoh,[13] and McMahon and Tincher[14] have reported similar findings for *meso* and *racemic* 2,4-dichloropentanes, and have given a much fuller interpretation. Figure I-4 shows the spectra obtained by the last authors. The *racemic* isomer's methylene resonance is a doublet of doublets. This is interpreted as indicating two equivalent protons, each of which exhibits two different couplings to the CH protons. This observation carries important conformational implications and we shall revert to it in Chapter II. The *meso* isomer shows for

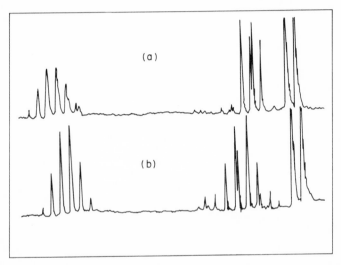

Fig. I-4. Spectra of (a) *racemic* and (b) *meso* 2,4-dichloropentane near room temperature; 20% (v/v) in CCl$_4$, 60 MHz [from P. E. McMahon and W. C. Tincher, *J. Mol. Spectry.*, **15**, 180 (1965)].

the methylene protons a more complex spectrum (approximately ABX$_2$) arising from strongly geminally coupled nonequivalent protons with about equal vicinal couplings. Analogous results have been reported for bromo-,[14] cyano-,[14,15,16] hydroxy-,[14,17] acetoxy-,[12,17] carboxy-,[18] carbomethoxy-,[19] and phenyl-[18,20] pentanes.

Such studies have been extended to the 2,4,6-trisubstituted heptanes, including the trichloro-,[21,22] tricarboxy-,[19] tricarbomethoxy-,[19] tricyano-,[23] and triphenyl-[24] heptanes. Of these, there are one *racemic* and two *meso* diastereoisomers, represented below in planar zigzag form, as seen along the zigzag plane:

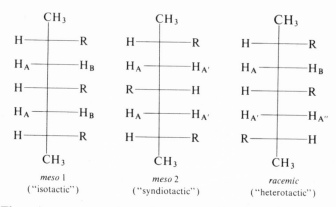

meso 1	*meso* 2	*racemic*
("isotactic")	("syndiotactic")	("heterotactic")

There is no accepted naming system for these diastereoisomers, but they are, as indicated, closely related to isotactic, syndiotactic, and heterotactic triads in vinyl polymer chains. All the methylene groups are heterosteric. The *meso* isomers each have two like heterosteric methylene groups; the racemic isomer has two unlike heterosteric methylene groups, so that all four of its methylene protons are non-equivalent. One might expect that the non-equivalence of the A′ and A″ protons would be small, while that of the A and B protons might be quite large; this expectation appears to be borne out.[21-24]

It has been assumed in the foregoing discussion that the molecules may exist in any or all of their possible conformations and that the rate of interconversion of conformations is rapid on the NMR time scale. Thus, for large molecules (including polymers), as for simple ethanic rotors, the existence of heterostericity or homostericity does not in principle depend upon any particular conformation being energetically preferred. This point will be further elaborated in Chapter III.

2. *Polymer Spectra: Polymethyl Methacrylate.* It is now

well recognized that these symmetry considerations can be extended to the long chains of polymers. In one way, very long chains simplify matters. If the degree of polymerization exceeds about 100, so that end effects can be neglected, then for a purely syndiotactic chain, H_A and $H_{A'}$ can be considered strictly equivalent, and in the isotactic chain only one kind of CH_AH_B group needs to be recognized. Figure I-5 shows the 60 MHz spectra of (a) predominantly isotactic and (b) predominantly syndiotactic polymethyl methacrylate in chlorobenzene solution at 150°. The methylene resonance of (b) is approximately the expected singlet, although somewhat broadened and complicated by the residual isotactic resonance and by additional effects with which we shall deal shortly. The methylene resonance in (a) is predominantly the expected AB quartet ($J = -14.9$), with additional structure due to syndiotactic sequences. The nature of these multiplets is an absolute measure of the polymer's predominant configuration, and it would not be necessary to have recourse to X-ray diffraction, even if it were possible.

The spectrum of the α-substituents does not in itself provide an absolute indication of configuration, but must be correlated with the β-methylene resonance or some other absolute measurement. If this is done, the α-substituent spectrum may give more detailed configurational information than the β-methylene spectrum. The chemical shifts of α-substituents commonly vary appreciably with the relative configurations of the nearest neighboring monomer units, but this is not always the case. For example, the α-methyl group resonance in polymethyl methacrylate is markedly dependent on configuration (Fig. I-5), whereas the ester methyl resonance is not. When such discrimination is possible one can observe three species of α-substituents in a polymer which is not stereochemically pure: those on the central monomer units of isotactic, syndiotactic, and heterotactic triads of monomer units. The definitions of these, as is now well known, correspond to the configurations shown above for the 2,4,6-trisubstituted heptanes. The simplest notations for

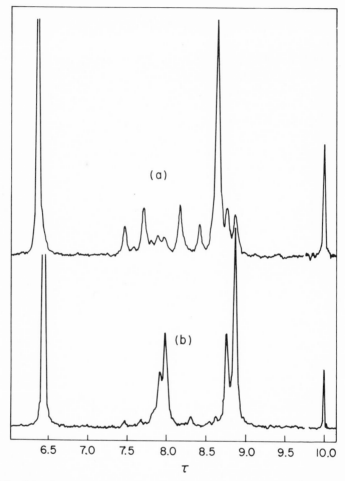

Fig. I-5. Spectra of 15% solutions in chlorobenzene of polymethyl methacrylate prepared with (*a*) an anionic initiator (PhMgBr) and (*b*) a free-radical initiator.

these triad sequences are *i*, *s*, and *h*, respectively.[25] A more general notation will be indicated a little later.

The designation of the configuration of a vinyl monomer unit in a polymer chain as *d* or *l*, although quite common, is open to the objection that it suggests distinctions which are not experimentally observable, since vinyl polymers ordinarily show no optical activity. A more serious objection to this notation is that it conflicts with the conventional use of the same notation to describe analogous small molecules. Thus, a methylene group in an isotactic chain can be regarded as the center unit in a *meso* dyad of monomer units having the same configuration, and such a dyad might be designated *dd* (or *ll*); but in the analogous small molecule, the 2,4-disubstituted pentane, the two asymmetric carbon atoms are of opposite configuration and are conventionally designated *d* and *l*. Similarly, for racemic dyads in syndiotactic chains, the monomer units are regarded by the polymer chemist as being of opposite configuration, whereas in the analogous pentane they have the same configuration. It is urged that the *d* and *l* notation be confined to small molecules and to polymers having true asymmetric centers (rather than pseudoasymmetric), such as polypeptides, polypropylene oxide, and vinyl polymers with asymmetric sidechains.

In Table I-1, the designations of monomer dyad and triad sequences are indicated. The *meso* dyad is designated *m* and the racemic dyad *r*. This system of nomenclature can be extended to sequences of any length. Thus, an isotactic triad is *mm*, a heterotactic triad *mr*, and a syndiotactic triad *rr*. Let us now assume that the probability of generating a *meso* sequence when a new monomer unit is formed at the end of a growing chain can be denoted by a single parameter, which we shall call P_m (this probability has been previously designated α^{26} and σ^{25}). In these terms, the generation of the chain is then a Bernoulli-trial process. In making this assumption, we need not be concerned with the details of the monomer addition step, but must assume that the probability of forming an *m* or *r* sequence is independent of the stereochemical configuration of the chain already formed. (This point will be more fully discussed in Chapter II.) It follows of course that the probability of forming an *r* sequence is $1 - P_m$. A triad sequence involves two monomer additions; the probabilities of *mm*, *mr*, and *rr* are (see column four of Table I)

TABLE I-1

		α-Substituent			β-CM₂		
	Designation	Projection	Bernoullian Probability		Designation	Projection	Bernoullian Probability
Triad	isotactic, mm (i) heterotactic, mr (h) syndiotactic, rr (s)		P_m^2 $2P_m(1-P_m)$ $(1-P_m)^2$	Dyad	meso, m racemic, r		P_m $(1-P_m)$
Pentad	mmmm (isotactic) mmmr rmmr mmrm mmrr rmrm (heterotactic) rmrr mrrm rrrm rrrr (syndiotactic)		P_m^4 $2P_m^3(1-P_m)$ $P_m^2(1-P_m)^2$ $2P_m^3(1-P_m)$ $2P_m^2(1-P_m)^2$ $2P_m^2(1-P_m)^2$ $2P_m(1-P_m)^3$ $P_m^2(1-P_m)^2$ $2P_m(1-P_m)^3$ $(1-P_m)^4$	Tetrad	mmm mmr rmr mrm rrm rrr		P_m^3 $2P_m^2(1-P_m)$ $P_m(1-P_m)^2$ $P_m^2(1-P_m)$ $2P_m(1-P_m)^2$ $(1-P_m)^3$

P_m^2, $2P_m(1 - P_m)$, and $(1 - P_m)^2$, respectively. A plot of these relations is shown in Fig. I-6. It will be noted that the proportion of *mr* units rises to a maximum at $P_m = 0.5$, corresponding to random propagation. For a random polymer, the proportion *mm*:*mr*:*rr* will be 1:2:1. (A similar plot of dyad frequencies against P_m would obviously be two straight lines with slopes of $+1$ and -1 for *m* and *r*, respectively.) For any given polymer, if Bernoullian, the *mm*, *mr*, *rr* sequence frequencies, as estimated from the relative areas of the appropriate peaks (α-methyl peaks for methyl methacrylate polymers), should lie on a single vertical line in Fig. I-6, cor-

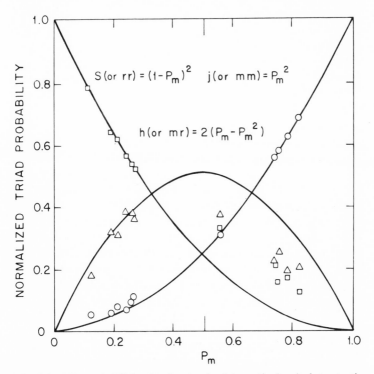

Fig. I-6. The probabilities (or fractions) of isotactic (*mm*), heterotactic (*mr*) and syndiotactic (*rr*) triads as a function of P_m, the probability of isotactic monomer placement during propagation.

responding to a single value of P_m. If this is not the case, then the polymer's configurational sequence deviates from Bernoullian. The polymer whose spectrum is shown in Fig. I-5a obeys these simple statistics within experimental error, P_m being 0.20 ± 0.01. The spectrum in Fig. I-5b shows that this polymer does *not* obey Bernoullian statistics, as we shall see in Chapter II (p. 53).

A considerable number of vinyl polymer spectra have now been observed and at least partially interpreted in these terms, including polypropylene,[27-31] polystyrene,[20,32] poly-α-methylstyrene,[33-35] polyvinyl chloride,[22,36-45] polyvinyl fluoride,[40,46] polytrifluorochloroethylene,[10] polyvinyl methyl ether,[40,47-49] poly-α-methylvinyl methyl ether,[50] polyvinyl acetate,[40,51-53] polyvinyl trifluoroacetate,[51] polyvinyl alcohol,[50,54-57] polyisopropyl acrylate,[58-60] polymethyl acrylate,[61,62] polymethyl methacrylate and other methacrylate esters,[24,63-76] polyvinyl formate,[77] polyacrylonitrile,[78-83] polymethacrylonitrile,[84] poly-2-vinylpyridine,[85] and polyacetaldehyde.[86-88]

As discrimination increases, through improved techniques and advances in instrumental design, finer structure becomes observable. With respect to β-methylene groups (or, more generally, β-CM$_2$ groups) one may expect to resolve *tetrad* sequences of monomer units, appearing as a fine structure on the *m* and *r* resonances. As indicated in Table I-1, the *m* resonance should be resolved into three tetrad resonances, all heterosteric, giving six different chemical shifts. The *r* resonance should be split into two homosteric resonances and one heterosteric resonance, giving a total of ten observable β-CM$_2$ chemical shifts in a polymer which is not too highly stereoregular. One may also expect the α-substituent resonance to be resolved into ten peaks corresponding to ten different *pentad* configurational sequences. If the polymer has been generated by a Bernoulli-trial propagation, the sequence frequencies will be as shown in columns four and eight of Table I-1. These relationships are plotted for tetrad sequences in Fig. I-7 and for pentad sequences in Fig. I-8. Since *m* and *r* are conjugate terms in these relationships, the frequency plots are symmetrical, as indicated in the legends to these figures: three curves

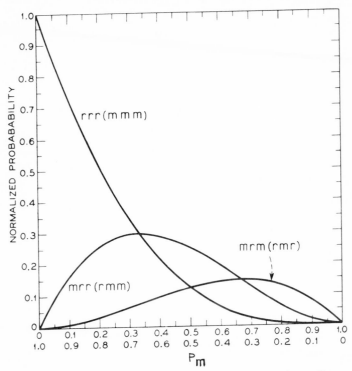

Fig. I-7.   Tetrad probabilities (or fractions) as a function of P_m.   For *rrr*, *mrr*, and *mrm* the upper P_m scale is used; for *mmm*, *rmm*, and *rmr* the lower P_m scale is used.

serve for six tetrad species and only four curves for ten pentad species, as there are four pairs of sequences of which the members of each pair have the same probability.

Since, as we shall see, these sequences apparently represent the limits of resolution attainable at present, we shall not describe longer sequences in detail. However, it may be instructive to point out[89] that if $N(n)$ is the number of distinguishable n (ads), we have the values

n	2	3	4	5	6	7	8
$N(n)$	2	3	6	10	20	36	72

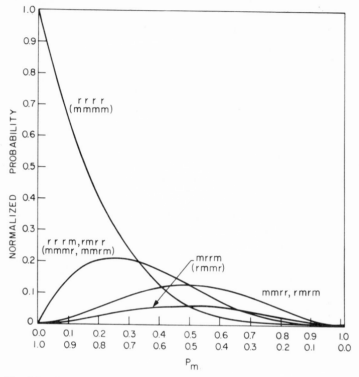

Fig. I-8. Pentad probabilities (or fractions) as a function of P_m. Use bottom scale for sequences in parentheses.

or, in general

$$N(n) = 2^{n-2} + 2^{m-1}$$

where $m = n/2$ if n is even and $m = (n - 1)/2$ if n is odd. It is evident that, regardless of the generating statistics, the number of possible types of sequences increases rapidly with their length (asymptotically as 2^{n-2} as $n \rightarrow \infty$), and that their discrimination beyond pentads, or hexads at most, will be very difficult.

It should be pointed out that the *direction* of the sequences is immaterial, except perhaps at the ends of the chain, which we ignore. Thus, for example, *mmr* cannot be observationally distinguished from *rmm*, nor can *rmrr* be distinguished from *rrmr*. Thus, "lopsided" sequences of this sort must be counted twice, *i.e.*, in both directions, and this is the reason that the factor 2 appears in their Bernoullian probability expressions (Table I-1). In polymers which have true asymmetric centers in the main chain, such as polypeptides or polypropylene oxide, a chain direction can be defined, and therefore, for example, the sequence *dldd* can (in principle, at least) be observationally distinguished from *dldd*.

The methylene spectrum of the predominantly syndiotactic polymethyl methacrylate (Fig. I-5*b*) shows a clear splitting (alluded to previously; see also ref. 76) into two peaks with a third weaker one as a shoulder on the low-field side. These same resonances can be seen, in altered relative intensity, in the spectrum of the predominantly isotactic polymer (Fig. I-5*a*). It is also significant that in the residual *meso* quartet in Fig. I-5*b*, the upfield doublet is at markedly higher field than the corresponding doublet in Fig. I-5*a*. These features clearly represent a discrimination of tetrad frequencies. It is well known that as the magnetic field (and observing r.f. field frequency) of the NMR spectrometer are increased, chemical shift differences, expressed on a gauss or frequency scale, increase proportionately, whereas spin-spin *J* couplings are unaffected. The result is improved discrimination and spectral simplification. The 100 MHz spectra of these polymers show an appreciable improvement over the 60 MHz results. A much more dramatic improvement in discrimination, however, is achieved with a 220 MHz instrument employing a 51.7 kilogauss superconducting solenoid.[90] The spectra in Figs. I-9*a* and I-9*b* are for the same polymers as in Fig. I-5; they were observed by Dr. R. C. Ferguson (duPont) and are reproduced with his permission. In Fig. I-9*a*, the spectra of the α-methyl groups (*ca.* 8.6–8.9τ) and β-methylene protons (*ca.* 7.4–8.6τ) are run at two amplifications; in Fig. I-9*b*, only the β-methylene protons are so shown. The resolution of tetrad signals is much clearer in these spectra, and a partial resolution of pentad signals is also achieved. The assignment

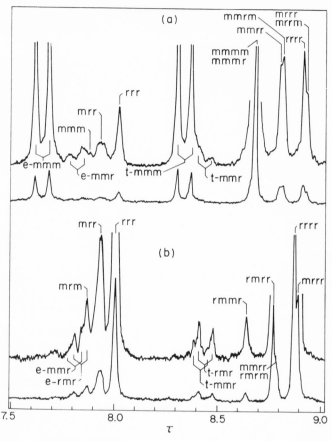

Fig. I-9. 220 MHz NMR spectra of polymethyl methacrylates. The polymers are the same as in Fig. I-5.

of peaks (indicated directly on the spectra) is based largely on the assumption that the predominantly syndiotactic polymer is Bernoullian, and the comparison of observed intensities with the appropriate curves of Figs. I-7 and I-8. In heterotactic methylene groups, it is desirable to make a definite assignment of the two distinguishable protons, as we shall see in Chapter II. This has been done for polyacrylate chains in a

very elegant study of model compounds by Yoshino *et al.*[91]
If we can assume that the same relative chemical shifts hold
for methacrylate chains as well, the assignment is

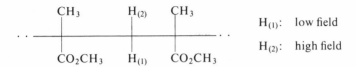

$H_{(1)}$: low field

$H_{(2)}$: high field

By an extension of the terminology used in describing small
molecules having two dissimilar asymmetric carbons, $H_{(1)}$ is
called the *erythro* proton and $H_{(2)}$ the *threo* proton. (The
basis of this terminology will be more evident when we discuss β-deuterium-labelled polymers in Chapter II.)

Peak assignments, particularly in the spectra of clearly non-
Bernoullian polymers such as in Fig. 1-9*a*, are also aided by
certain necessary relationships among the frequencies of occurrence of sequences, which must hold regardless of the configurational statistics. The most useful of these are shown in
Table I-2. These often allow seemingly tempting assignments

**TABLE I-2. Some Necessary Relations Among Sequence
Frequencies**

Dyad:	$(m) + (r) = 1$
Triad:	$(mm) + (mr) + (rr) = 1$
Dyad-Triad:	$(m) = (mm) + \frac{1}{2}(mr)$
	$(r) = (rr) + \frac{1}{2}(mr)$
Triad-Tetrad:	$(mm) = (mmm) + \frac{1}{2}(mmr)$
	$(mr) = (mmr) + 2(rmr) + (mrr) + 2(mrm)$
	$(rr) = (rrr) + \frac{1}{2}(mrr)$
Tetrad-Tetrad:	sum = 1
	$(mmr) + 2(rmr) = 2(mrm) + (mrr)$
Pentad-Pentad:	sum = 1
	$(mmmr) + 2(rmmr) = (mmrm) + (mmrr)$
	$(mrrr) + 2(mrrm) = (rmrr) + (rrmm)$
Tetrad-Pentad:	$(mmm) = (mmmm) + \frac{1}{2}(mmmr)$
	$(mmr) = (mmmr) + 2(rmmr) = (mmrm) + (mmrr)$
	$(rmr) = \frac{1}{2}(mrmr) + \frac{1}{2}(rmrr)$
	$(mrm) = \frac{1}{2}(mrmr) + \frac{1}{2}(mmrm)$
	$(rrm) = 2(mrrm) + (mrrr) = (mmrr) + (rmrr)$
	$(rrr) = (rrrr) + \frac{1}{2}(mrrr)$

to be ruled out, and permit choices between otherwise equally plausible alternatives.

It is noteworthy that the positions of the tetrad resonances are the same within experimental error in the predominantly isotactic and predominantly syndiotactic polymers, despite what may be assumed to be different conformations of the chain. The pentad resonances are shifted slightly but observably (*ca.* 0.02τ) upfield in the isotactic polymer. The α-methyl group chemical shifts are thus somewhat more sensitive to changes in the predominant conformation of the chain. However, these observations do not really constitute a test of the prediction of Flory and Baldeschwieler[92] of rather large effects of this sort, as the methacrylate chain is probably not, even locally, in the 3_1-helical conformation upon which their calculations are based (see Chapter III).

3. *The Effect of Polymerization Temperature.* The effect of the temperature of free-radical polymerization upon the configuration of the resulting polymer chain is of considerable interest, and is amenable to study by NMR. If we express the propagation steps in terms of absolute reaction-rate theory, we have for the rate constant for isotactic propagation

$$k_i = (kT/h)\exp\{(\Delta S_i^{\ddagger}/R) - (\Delta H_i^{\ddagger}/RT)\} \qquad (I\text{-}1)$$

and for syndiotactic propagation

$$k_s = (kT/h)\exp\{(\Delta S_s^{\ddagger}/R) - (\Delta H_s^{\ddagger}/RT)\} \qquad (I\text{-}2)$$

from which

$$P_m = k_i/(k_i + k_s)$$

$$= \frac{\exp\{-(\Delta G_i^{\ddagger} - \Delta G_s^{\ddagger})/RT\}}{1 + \exp\{-(\Delta G_i^{\ddagger} - \Delta G_s^{\ddagger})/RT\}} \qquad (I\text{-}3)$$

and

$$P_m/(1 - P_m) = k_i/k_s$$

$$= \exp\{(\Delta S_i^{\ddagger} - \Delta S_s^{\ddagger})/R - (\Delta H_i^{\ddagger} - \Delta H_s^{\ddagger})/RT\}$$

$$(I\text{-}4)$$

We therefore have for the difference in activation enthalpies for propagation

$$\Delta(\Delta H_p^{\ddagger}) = \Delta H_i^{\ddagger} - \Delta H_s^{\ddagger}$$
$$= -R \, \partial \ln[P_m/(1 - P_m)]/\partial(1/T) \qquad \text{(I-5)}$$

and for the difference in activation entropies

$$\Delta(\Delta S_p^{\ddagger}) = \Delta S_i^{\ddagger} - \Delta S_s^{\ddagger}$$
$$= R \ln[P_m/(1 - P_m)] + \Delta(\Delta H_p)/T \qquad \text{(I-6)}$$

The structures of polymers prepared over a 178° temperature range are summarized in Table I-3. A plot of these data

TABLE I-3.   The Stereochemical Configuration of Free Radical
Polymethyl Methacrylate as a Function of
Polymerization Temperature

Temp., °C	$[\eta]^{25°}$ (benzene)	$M_w \times 10^{-3}$	(mm)	(mr)	(rr)	P_m	$\dfrac{P_m}{(1 - P_m)}$
−78	0.10	19	0.04_8	0.17_2	0.78_0	0.12	0.13_6
0	0.58	200	0.07_5	0.30_0	0.62_5	0.21	0.26_6
50	0.27	69	0.08_5	0.31_5	0.60_0	0.23	0.30_0
100	0.46	140	0.08_9	0.37_5	0.53_9	0.27	0.37_0

according to eq. (I-4) is shown in Fig. I-10. From the slope of this Arrhenius plot, values of 775 ± 75 cal. for $\Delta(\Delta H_p^{\ddagger})$ and 0.0 ± 0.1 e.u. for $\Delta(\Delta S_p^{\ddagger})$ are obtained. Redetermination of these quantities[66,74] indicates that more accurate values are *ca.* 1 kcal. and 1 e.u., respectively. Thus, the preference for syndiotactic placement is due to a small additional energy required for isotactic placement; the latter is favored by entropy.

4. *Polyvinyl Chloride.* The spectrum of polymethyl methacrylate is relatively simple because, although there is a strong geminal coupling of the protons of the *meso* methylene groups, there is no observable coupling of the methylene and α-methyl protons. In polymers of monosubstituted

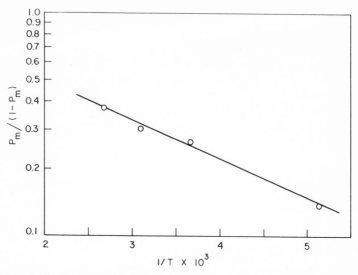

Fig. I-10. Arrhenius plot of the stereoregularity of polymethyl methacrylate on T, using free-radical initiation.

monomers, such as vinyl chloride, the *vicinal* coupling of the α- and β-methylene protons produces a splitting of the observed resonances. On the simplest basis one would expect the α-protons, being coupled to four neighboring β-protons, to appear as a quintuplet and the β-protons, coupled to two α-protons, as a triplet. The observed spectrum (Fig. I-11) does indeed apparently show a quintuplet centered at 5.53τ (in chlorobenzene solution) and a group of five peaks centered at about 7.9τ for the methylene protons. The methylene resonance is essentially two overlapping triplets centered at 7.78τ and 7.96τ, corresponding to *meso* and *racemic* methylene groups, respectively[36].

This interpretation is confirmed by *double resonance*. In this technique, one multiplet corresponding to a group of coupled spins is observed using the usual weak radio-frequency field while simultaneously a second multiplet, corresponding to another nucleus or group of nuclei coupled to the first group, is irradiated with a much stronger radio-frequency field. This has the effect of abolishing the coupling between the spins. For optimum effect, the difference in frequency between the two fields, $\Delta\nu$, must correspond to the chemical shift difference between the two groups of spins. As shown in Fig. I-11, when the β-protons are decoupled from the α-protons ($\Delta\nu = 137$ cps), two peaks are observed,[40] separated by 0.20 p.p.m. [Fig. I-11 (*b*)]. A very similar spectrum is shown by the polymer of α-deuterovinyl chloride:

$$\begin{array}{c} Cl \\ \diagdown \\ \diagup \\ D \end{array} C{=}CH_2$$

The substitution of deuterium for hydrogen represents another means of spectral simplification, for the resonance of deuterium is very far removed from that of hydrogen; in addition, the H-D coupling has less than one-sixth the magnitude of the corresponding H-H coupling, and produces no observable multiplicity in polymer spectra [Fig. I-11 (*d*)].

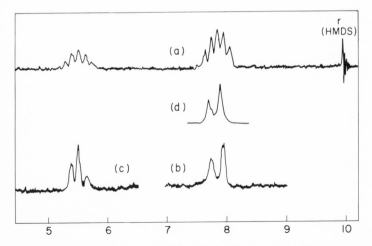

Fig. I-11. Normal and decoupled spectra of polyvinyl chloride and poly-α-d_1-vinyl chloride (see text).

When the CH_2 protons are irradiated ($\Delta\tau = 133$ cps), the α-proton resonances show peaks at 5.48τ, 5.59τ, and 5.71τ [Fig. I-11(c)]; the dependence of the decoupled α-proton spectrum upon $\Delta\tau$ has been employed[40] to demonstrate that these peaks correspond to syndiotactic, heterotactic, and isotactic triads, respectively, assuming the correctness of the above assignment of the β-proton resonances. The normal α-proton spectrum thus is actually three overlapping quintuplets.

Yoshino and Komiyama[42] have shown that the spectrum of poly-α-cis-β-d_2-vinyl chloride exhibits the ten different chemical shifts expected if one can discriminate all six of the β-methylene tetrad resonances. It is observed that the central β-protons in all three of the *meso* tetrads are heterosteric, as expected (Table I-1) but the nonequivalence is substantial only for the *rmr* tetrad. The presence of geminal coupling, as in normal polyvinyl chloride chloride or poly-α-d_1-vinyl chloride (Fig. I-11), tends to concentrate intensity at the center of the band, and this effect is accentuated by the fact that the *mean* chemical shift for each heterosteric pair of protons is nearly the same for all *meso* tetrads.

In these terms, the spectrum of polyvinyl chloride can be quite fully interpreted. In Fig. I-12a the spectrum shown in Fig. I-11 (d) is presented in greater detail. This polymer was prepared at 100°. Curve c shows calculated AB quartets (for heterosteric CH_2 groups) and singlets (for homosteric CH_2 groups) corresponding to the six methylene tetrads. The assignments of the chemical shifts are shown in this figure; these assignments are the same for the other three polymers, prepared at 50°, 0°, and −78° (Figs. I-12 and -13). The relative intensities of the calculated resonance lines are obtained from Fig. I-7—*i.e.*, from Bernoullian statistics—choosing values of P_m which give the best match of experimental and calculated spectra at each temperature. The calculated spectrum (b) in Fig. I-12, like the corresponding calculated spectra for the other three polymers, is obtained from a computer program which allows one to introduce any chosen linewidth (2.5 cps in these spectra) and to sum up the

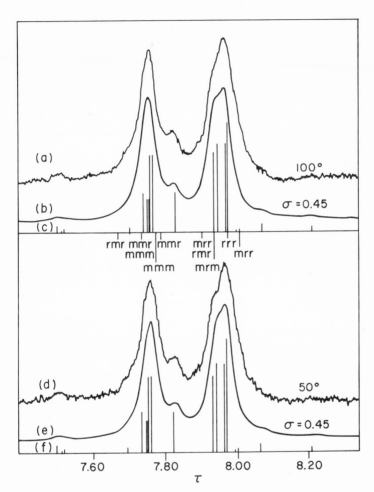

Fig. I-12. Calculated and experimental NMR spectra of poly-α-d_1-vinyl chloride prepared at varying temperatures of polymerization. (*a*) Observed spectrum of 100° polymer; (*b*) calculated spectrum of 100° polymer, with linewidth of 2.5 cps; (*c*) "stick" spectrum corresponding to (*b*); (*d*) observed spectrum of 50° polymer; (*e*) calculated spectrum of 50° polymer with linewidth of 2.5 cps; (*f*) "stick" spectrum corresponding to (*e*). On the abscissa of spectra (*a*), (*b*), and (*c*) are shown the chemical shifts of the six tetrads (see text).

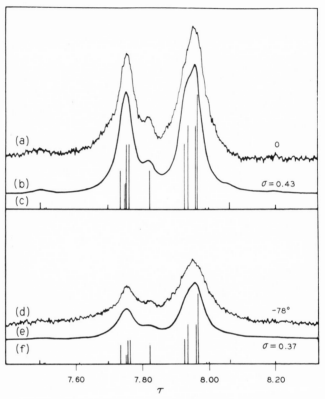

Fig. I-13. NMR spectra of poly-α-d_1-vinyl chloride. (*a*) Observed spectrum of polymer prepared at 0°; (*b*) calculated spectrum of the 0° polymer, with linewidth of 2.5 cps; (*c*) "stick" spectrum corresponding to (*b*); (*d*) observed spectrum of −78° polymer; (*e*) calculated spectrum of −78° polymer, with linewidth of 3.2 cps; (*f*) "stick" spectrum corresponding to (*e*).

resulting spectra in any desired proportion. By visual inspection and matching of relative peak heights in calculated and experimental spectra, the following values of P_m are deduced: −78°:0.37; 0°:0.43; 50°:0.45; 100°:0.46. From these data and eqs. (I-5) and (I-6), it is found that

$$\Delta(\Delta H_p)^\ddagger = 310 \pm 20 \text{ cal}$$
$$\Delta(\Delta S_p)^\ddagger = 0.6 \pm 0.1 \text{ e.u.}$$

There is thus a measurable tendency for vinyl chloride to give more syndiotactic polymers at lower temperatures, but the tendency is much smaller than for methyl methacrylate and appears hardly sufficient to account for the alteration in properties observed for the low-temperature polymers, particularly the enhanced crystallizability.[93]

5. *Polystyrene.* The earliest reported polymer spectra were those of polystyrene,[94,95] which in carbon tetrachloride solution gave an aromatic resonance showing a separate peak for the *ortho* protons, upfield from the *meta-para* proton resonance, and a single broad peak for the aliphatic protons. In Fig. I-14, curve *d* shows the 60 MHz spectrum of isotactic polystyrene obtained using 15% solutions in tetrachloroethylene at 128° for the aromatic protons and in *o*-dichlorobenzene at 200° for the main-chain protons.[20] Since there is no significant coupling of ring and main-chain protons, they may be treated as separate systems. Such "strong-coupled" spectra, *i.e.*, where the chemical-shift differences are comparable in magnitude to the couplings between the protons, cannot be solved for values of τ and J by inspection, since the spacings and line positions do not provide these parameters directly. Instead, quantum mechanical calculations are necessary. In the case of a polymer, some simplified spin model is also necessary, since one obviously cannot include *all* the protons present in a long chain. Following the suggestion of Tincher,[31,39] the polystyrene backbone protons are treated as if the chain were only a cyclic dimer. In Fig. I-14, curve *f* shows the line spectrum resulting from a six-spin "dimer" calculation, using the parameters given in the figure title. The results of a five-spin calculation for the aromatic protons are also shown. Curve *e* shows the calculated spectrum with the experimental linewidths included. It can be seen that the matching is satisfactory. From this result one can draw the following conclusions:

1. The phenyl spin system is symmetrical about a bisecting plane through the *para* carbon atom, and thus gives no evidence of a high (*i.e.*, exceeding *ca.* 15 kcal) barrier to rotation.

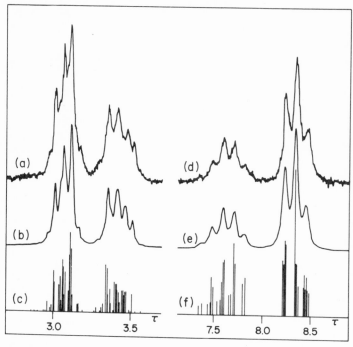

Fig. I-14. Isotactic polystyrene (*a*) Aromatic proton spectrum, in tetra-chloroethylene at 135°. (*b*) Calculated spectrum $\nu(ortho) - \nu(meta) = 22$ cps; $\nu(para) - \nu(meta) = 2$ cps; all $J(ortho) = 6.5$ cps; both $J(meta) = 2.0$ cps; $J(para) = 1.0$ cps; linewidth = 1.0 cps. (*c*) "Stick spectrum" corresponding to (*b*). (*d*) Backbone-proton spectrum in *o*-dichlorobenzene at 200°. (*e*) Calculated spectrum: $\nu_\beta - \nu_\alpha = 43.8$ cps; both $J(rac) = 7.0$ cps; $J(gem) = 14.0$ cps; linewidth = 3.0 cps. (*f*) "Stick spectrum" corresponding to (*e*).

2. The two *meso* β-protons are fortuitously equivalent in chemical shift within experimental error (± 0.05 ppm).
3. The two β-protons show couplings (7.00 ± 0.20 cps) to the α-protons which are equal within experimental error. This observation is consistent with the assumption that the polymer chain has a *local* conformation in solution which is the same as in the 3_1 helix which it is

known to have in the crystal. (This will be more fully discussed in Chapter III.)

"Atactic" polystyrene, such as is normally obtained by free-radical and ionic polymerization, gives a much less satisfactory spectrum.[94-96] Brownstein et al.[32] have observed that in the α-proton spectrum of "atactic" poly-$\beta\beta$-d_2-styrene, resonances corresponding to isotactic, heterotactic, and syndiotactic triads can be distinguished, and this observation has been confirmed.[20] Figures I-15a and I-15b show the spectra for "atactic" polystyrene, prepared by free-radical polymeri-

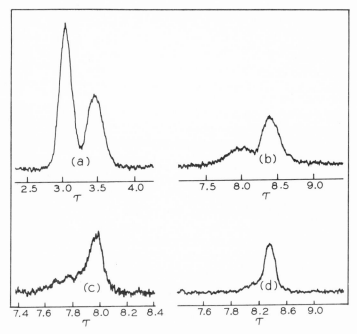

Fig. I-15. "Atactic" polystyrene. (a) Aromatic-proton spectrum, in tetrachloroethylene at 135°. (b) Backbone-proton spectrum, in o-dichlorobenzene at 200°. (c) Poly (β,β-d_2-styrene) in o-dichlorobenzene at 200° with deuterium irradiation. (d) Poly-α-d_1-styrene in o-dichlorobenzene at 200° with deuterium irradiation.

zation, in tetrachloroethylene at 135° and in *o*-dichloroben-
zene at 200°. Figure I-15*c* shows the α-proton spectrum of
"atactic" poly-β-d_2-styrene and Fig. I-15*d* shows the β-proton
spectrum of "atactic" poly-α-d_1-styrene, both prepared in the
same way and observed at 200° in *o*-dichlorobenzene. Three
peaks can be discerned in the α-proton spectrum. The peak at
lowest field (7.68τ) is in approximately the same position (al-
lowing for the increased shielding which is known to be produced by neighboring deuterium nuclei) as that for isotactic
polystyrene (7.61τ), indicating an assignment of the predominant peak to syndiotactic triads. These polymers appear to
be about 80% syndiotactic. The β-proton spectrum shows
only a broad singlet, the *meso* protons not being distinguishable from each other (as expected from the foregoing discussion) or from the much larger *racemic* singlet.

The atactic polymer spectra always appear to be much
more poorly resolved than the isotactic ones. This is to be
attributed to the overlapping of small but appreciable chemical-shift differences of pentads and tetrads.

6. *Polyisopropyl Acrylate.* The spectrum of isotactic polyisopropyl acrylate has been reasonably satisfactorily solved;[58]
the experimental spectrum of the backbone protons is shown
in Fig. I-16, together with calculated spectra obtained using
the same "dimer" model as employed for isotactic polystyrene. The NMR parameters employed are shown in the title
of Fig. I-16. It will be seen that, as in isotactic polymethyl
methacrylate and in distinction to isotactic polystyrene, the
methylene protons show a substantial difference in chemical
shift. Our calculations are made on the assumption that the
couplings of the α-protons are equal to both β-methylene
protons. Recently, however, Yoshino *et al.*[60] have shown
that the spectrum of polyisopropyl β-d_1-acrylate is not consistent with this assignment, and shows a greater coupling of
the B than of the A β-methylene protons. This observation
has an important bearing on the question of the conformation of the polymer chain in solution (Chapter III).

The isotactic polyacrylate backbone spectrum represents an

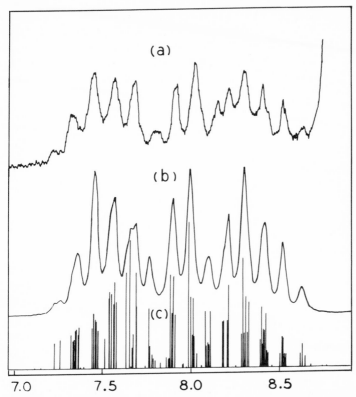

Fig. I-16. Spectra of isotactic poly(isopropyl acrylate). (*a*) Observed in chlorobenzene at 150°; (*b*) calculated spectrum with 2.5 cps linewidth; (*c*) calculated "stick" spectrum, both with the following parameters: CH: 7.43τ; CH_2: $H_A = 7.86\tau$, $H_B = 8.32\tau$; J (*gem*): -13.6 cps; both J (*vic*): 6.50 cps.

even more strongly coupled spin system than that of isotactic polystyrene, and the computer simulation is less satisfactory. For example, intensity deviations can be observed in Fig. I-16. These probably mean that in this case the "dimer" model is too simple. It is a consequence of this (and of the generally inferior resolution of polymer backbone proton spectra) that there is a considerable uncertainty in the calculated J-couplings. It is thus very possible that Yoshino's couplings are the more accurate.*

7. *Polypropylene.* A further step in complexity is presented by polypropylene, in which the coupling of sidechain protons to the main-chain α-protons, not heretofore observable, strongly affects the spectrum. The first useful results[27] were obtained by employing the deutero-monomer

$$
\begin{array}{c}
CD_3 \\
| \\
C{=}CH_2 \\
| \\
D
\end{array}
$$

Solvent fractionation of a polymer prepared with a Ziegler-Natta initiator from this monomer gave the results shown in Fig. I-17. The crystallizable fraction insoluble in ethyl ether gives an AB quartet for the methylene resonance, and appears thus to be almost purely isotactic. The most soluble fraction gives largely a singlet, and is predominantly syndiotactic. There was also obtained what appeared to be a "block" isotactic-syndiotactic fraction (see below), although this could not be demonstrated from the CH_2 spectrum alone.

The spectrum of the undeuterated polymer has also yielded to analysis.[28-31] Figure I-18 shows[28] the 60 MHz spectra of isotactic (a) and syndiotactic (b) polypropylene, observed in o-dichlorobenzene at 150°. Eight broad peaks, not all clearly visible, centered at *ca.* 8.50τ are given by the α-protons, split by three methyl and four methylene protons. Centered at *ca.* 8.94τ in spectrum (b) is the distorted triplet of the methylene protons. This region is of course much more complex in spectrum (a) and overlaps the α-proton resonance. More useful for analytical purposes are the strong resonances of the methyl groups, split to doublets by the α-protons. In the isotactic spectrum this doublet appears at 9.13τ and in

---

*NOTE ADDED IN PROOF: More recent studies at 100 MHz [F. Heatley and F. A. Bovey, *Macromolecules*, **1**, 303 (1968)] have provided more precise values for these couplings: 6.0 and 7.5 cps, the larger coupling being to the less shielded β-proton. See also page 92 *et seq.*

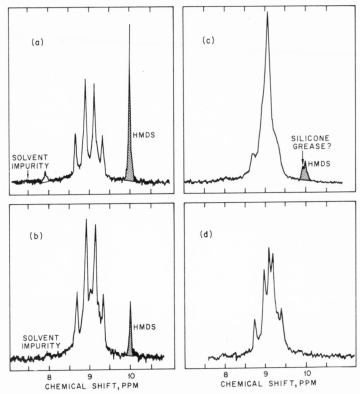

Fig. I-17. 60 MHz spectra of poly-2,3,3,3-d_4-propylene in 2-chlorothio-phene at 110°C. (*a*) Isotactic fraction; (*b*) stereoblock fraction; (*c*) amorphous fraction; (*d*) unfractionated fraction. Internal reference is hexa-methyldisiloxane ("HMDS") at 9.94τ. [From F. C. Stehling, *J. Polymer Sci.*, **A2**, 1815 (1964).]

the syndiotactic spectrum at 9.18τ. Figure I-19 shows the 100 MHz spectrum obtained for a polymer containing both isotactic and syndiotactic sequences, as indicated by two over-lapping methyl doublets. At an intermediate position, as expected, there appears the weak doublet of the heterotactic (*mr*) sequences, representing the junctions of the isotactic and syndiotactic "blocks" (Chapter II). The polymer thus de-

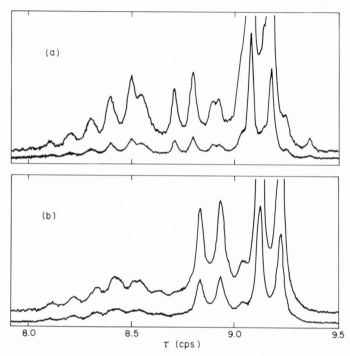

Fig. I-18. 60 MHz spectra of polypropylene in *o*-dichlorobenzene at 150°C; (*a*) isotactic; (*b*) syndiotactic. [From J. C. Woodbrey, *J. Polymer Sci.*, **B2**, 315 (1964).]

parts very far from Bernoulli-trial structure. Such block chains will in general be formed if any stereochemical configuration, once formed, tends to propagate itself preferentially but not exclusively. If we assign lengths to the blocks in the manner illustrated by this example:[25]

$$\underbrace{\ldots iiiii}_{\bar{n}_i} \quad \underbrace{hsssssss}_{\bar{n}_s} \quad hiiii \ldots$$

then

$$(\bar{n}_i - 1)/(\bar{n}_s - 1) = (mm)/(rr) \tag{I-7}$$

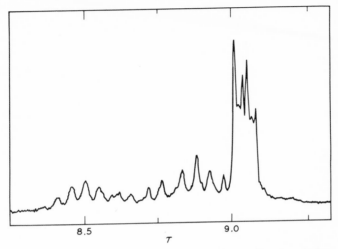

Fig. I-19.   100 MHz spectrum of stereoblock polypropylene in o-dichloro-benzene at 150°C. [From J. C. Woodbrey, *J. Polymer Sci.*, **B2**, 315 (1964).]

$$\bar{n}_i / \bar{n}_s = (m)/(r) \tag{I-8}$$

$$\bar{n}_i / (\bar{n}_i + \bar{n}_s) = (m)/[(r) + (m)] \tag{I-9}$$

$$(\bar{n}_i + \bar{n}_s)/2 = 1/(mr) = \bar{n} \tag{I-10}$$

where $\bar{n}_i$ and $\bar{n}_s$ are number-average lengths of the isotactic and syndiotactic blocks, respectively, and $\bar{n}$ is the number-average length of all blocks. [If the propagation is Bernoulli-trial, then $\bar{n}_i/\bar{n}_s = (m)/(r) = P_m/(1 - P_m)$, $\bar{n}_s = 1/P_m$, and $\bar{n}_i = 1/(1 - P_m)$.] Thus, a weak mr peak means that block lengths must be relatively long. For the polypropylene of Fig. I-19, $\bar{n}$ can be estimated from eq. (I-10) as about 16.

The better resolution in the 100 MHz spectrum is evident. In fact, the heterotactic doublet cannot be distinguished at 60 MHz. As might be expected, a still more striking spectral simplification is achieved at 220 MHz,[97] as shown in Fig. I-20, in which the α, β, and methyl protons are clearly separated, and the spectrum is nearly first-order.

These five examples have been chosen to illustrate the principles of the application of NMR to the measurement of

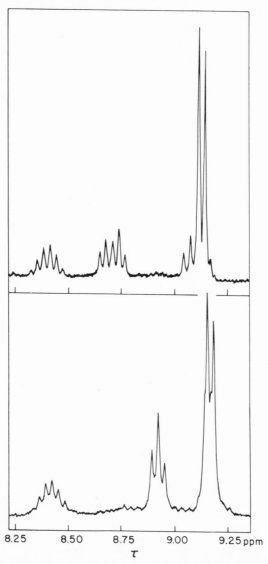

Fig. I-20. 220 MHz spectra of polypropylene in *o*-dichlorobenzene at 165°; (*a*) isotactic; (*b*) syndiotactic. (Reproduced by permission of Dr. R. C. Ferguson.)

chain configuration. In the next chapter, we shall consider in more detail the statistics of configurational sequences and their relationship to propagation mechanisms.

REFERENCES FOR CHAPTER I

(1) H. S. Gutowsky, *J. Chem. Phys.*, **37**, 2916 (1962), and references cited therein.

(2) J. S. Waugh and F. A. Cotton, *J. Phys. Chem.*, **65**, 562 (1961).

(3) H. S. Finegold, *Proc. Chem. Soc.*, **1960**, 283.

(4) P. R. Shafer, D. R. Davis, M. Vogel, K. Nagarajan, and J. D. Roberts, *Proc. Natl. Acad. Sci. U.S.*, **47**, 49 (1961).

(5) J. G. Prichard and P. C. Lauterbur, *J. Am. Chem. Soc.*, **83**, 2105 (1961).

(6) F. Kaplan and J. D. Roberts, *J. Am. Chem. Soc.*, **83**, 4666 (1961).

(7) T. D. Coyle and F. G. A. Stone, *J. Am. Chem. Soc.*, **83**, 4138 (1961).

(8) J. W. Wilt and W. J. Wagner, *Chem. and Ind.*, **1964**, 1389.

(9) G. C. Brumlik, R. L. Baumgarten, and A. J. Kosak, *Nature*, **201**, 388 (1964).

(10) G. V. D. Tiers and F. A. Bovey, *J. Polymer Sci.*, **A1**, 833 (1963).

(11) D. Doskocilova, *J. Polymer Sci.*, **B2** ("Polymer Letters"), 421 (1964).

(12) D. Doskocilova and B. Schneider, *Coll. Czech. Chem. Communs.*, **29**, 2290 (1964).

(13) S. Satoh, *J. Polymer Sci.*, **A2**, 5221 (1964).

(14) P. E. McMahon and W. C. Tincher, *J. Mol. Spectry.*, **15**, 180 (1965).

(15) H. G. Clark, *Makromol. Chem.*, **63**, 69 (1963).

(16) R. Yamadera and M. Murano, *J. Polymer Sci.*, **A1**, 1855 (1967).

(17) Y. Fujiwara and S. Fujiwara, *Bull. Chem. Soc. Japan*, **37**, 1005 (1964).

(18) D. Doskocilova and B. Schneider, *J. Polymer Sci.*, **B3**, 209 (1965).

(19) H. G. Clark, *Makromol. Chem.*, **86**, 107 (1965).

(20) F. A. Bovey, F. P. Hood, E. W. Anderson, and L. C. Snyder, *J. Chem. Phys.*, **42**, 3900 (1965).

(21a) T. Shimanouchi, M. Tasumi, and Y. Abe, *Makromol. Chem.*, **86**, 43 (1965).

(21b) Y. Abe, M. Tasumi, T. Shimanouchi, S. Satoh, and R. Chujo, *J. Polymer Sci.*, **4A1**, 1413 (1966).

(22) D. Doskocilova, J. Stokr, B. Schneider, H. Pivcova, M. Kolinsky, J. Petranek, and D. Lim, *J. Polymer Sci.*, **C16**, 215 (1967).

(23) M. Murano and R. Yamadera, *J. Polymer Sci.*, **B5**, 483 (1967).

(24) H. Pivcova, M. Kolinsky, D. Lim, and B. Schneider, Preprints IUPAC Macromolecular Symposium, Brussels, 1967.

(25) F. A. Bovey and G. V. D. Tiers, *J. Polymer Sci.*, **44**, 173 (1960).

(26) B. Coleman, *J. Polymer Sci.*, **31**, 155 (1958).

(27) F. C. Stehling, *J. Polymer Sci.*, **A2**, 1815 (1964).

(28) J. C. Woodbrey, *J. Polymer Sci.*, **B2**, 315 (1964).

(29) Y. Kato, N. Ashikari, and A. Nishioka, *Bull. Chem. Soc. Japan,* **37,** 1630 (1964).

(30) G. Natta, E. Lombardi, A. L. Segre, A. Zambelli, and A. Marinangeli, *Chim. e Ind.,* **47,** 378 (1965).

(31) W. C. Tincher, *Makromol. Chem.,* **85,** 20 (1965).

(32) S. Brownstein, S. Bywater, and D. J. Worsfold, *J. Phys. Chem.,* **66,** 2067 (1962).

(33) S. Brownstein, S. Bywater, and D. J. Worsfold, *Makromol. Chem.,* **48,** 127 (1961).

(34) Y. Sakurada, M. Matsumoto, K. Imai, A. Nishioka, and Y. Kato, *J. Polymer Sci.,* **B1,** 633 (1963).

(35) K. C. Ramey and G. L. Statton, *Makromol. Chem.,* **85,** 287 (1965).

(36) U. Johnsen, *J. Polymer Sci.,* **54,** 56 (1961).

(37) F. A. Bovey and G. V. D. Tiers, *Chem. Ind. (London),* **1962,** 1826.

(38) R. Chujo, S. Satoh, T. Ozeki, and E. Nagai, *J. Polymer Sci.,* **61,** 512 (1962).

(39) W. C. Tincher, *J. Polymer Sci.,* **62,** S148 (1962); *Makromol. Chem.,* **85,** 20 (1965).

(40) F. A. Bovey, E. W. Anderson, D. C. Douglass, and J. A. Manson, *J. Chem. Phys.,* **39,** 1199 (1963).

(41) S. Satoh, *J. Polymer Sci.,* **A2,** 5221 (1964).

(42) T. Yoshino and J. Komiyama, *J. Polymer Sci.,* **B3,** 311 (1965).

(43) O. C. Bockman, *ibid.,* **A3,** 3399 (1965).

(44) K. C. Ramey, *J. Phys. Chem.,* **70,** 2525 (1966).

(45) F. A. Bovey, F. P. Hood, E. W. Anderson, and R. L. Kornegay, *ibid.,* **71,** 312 (1967).

(46) C. W. Wilson, Paper presented to Division of Polymer Chemistry, 148th Meeting of American Chemical Society, Chicago, September, 1964.

(47) K. C. Ramey, N. D. Field, and I. Hasegawa, *J. Polymer Sci.,* **B2,** 865 (1964).

(48) S. Brownstein and D. M. Wiles, *J. Polymer Sci.,* **A2,** 1901 (1964).

(49) Y. Ohsumi, T. Higashimura, and S. Okamura, *J. Polymer Sci.,* **5A1,** 849 (1967).

(50) M. Goodman and Y. Fan, *J. Am. Chem. Soc.,* **86,** 4922 (1964).

(51) K. C. Ramey and N. D. Field, *J. Polymer Sci.,* **B2,** 63, 69 (1965); **B5,** 39 (1967).

(52) S. Murahashi, S. Nozakura, M. Sumi, H. Yuki, and K. Hatada, *ibid.,* **B4,** 65 (1966).

(53) K. Fujii, Y. Fujiwara, and S. Fujiwara, *Makromol. Chem.,* **89,** 278 (1965).

(54) A. Danno and N. Hayakawa, *Bull. Chem. Soc. Japan,* **35,** 1748 (1962).

(55) W. C. Tincher, *Makromol. Chem.,* **85,** 42 (1965).

(56) J. Bargon, D. H. Hellwege, and U. Johnsen, *Makromol. Chem.,* **85,** 291 (1965).

(57) S. Murahashi, S. Nozakura, M. Sumi, and K. Matsumura, *ibid.,* **B4,** 59 (1966).

(58) C. Schuerch, W. Fowells, A. Yamada, F. A. Bovey, F. P. Hood, and E. W. Anderson, *J. Am. Chem. Soc.,* **86,** 4481 (1964).

(59) A. Ueno and C. Schuerch, *J. Polymer Sci.,* **B3,** 53 (1965).

(60) T. Yoshino, Y. Kikuchi, and J. Komiyama, *J. Phys. Chem.,* **70,** 1059 (1966).

(61) K. Matsuzaki, T. Uryu, A. Ishida, and T. Oki, *J. Polymer Sci.,* **B2,** 1139 (1964).

(62) T. Yoshino, J. Komiyama, and M. Shinomiya, *J. Am. Chem. Soc.,* **86,** 4482 (1964).

(63) F. A. Bovey, *J. Polymer Sci.,* **46,** 59 (1960).

(64) U. Johnsen and K. Tessmar, *Kolloid-Z.,* **168,** 160 (1960).

(65) A. Nishioka, Y. Kato, T. Uetake, and H. Watanabe, *J. Polymer Sci.,* **61,** S32 (1962).

(66) T. G. Fox and H. W. Schnecko, *Polymer,* **3,** 575 (1962).

(67) D. W. McCall and W. P. Slichter, in "Newer Methods of Polymer Characterization," ed. by B. Ke, Interscience Publishers, New York, 1964, p. 231.

(68) K. Matsuzaki and S. Nakano, *J. Polymer Sci.,* **A2,** 3339 (1964).

(69) Y. Kato and A. Nishioka, *J. Polymer Sci.,* **B3,** 739 (1965).

(70) K. Yokota and Y. Ishii, *J. Polymer Sci.,* **B3,** 777 (1965).

(71) D. M. Wiles and S. Brownstein, *J. Polymer Sci.,* **B3,** 951 (1965).

(72) Y. Kato and A. Nishioka, *J. Chem. Soc. Japan,* **68,** 1461 (1965).

(73) T. Tsuruta, T. Makimoto, and H. Kanai, *J. Macromol. Chem.,* **1,** 31 (1966).

(74) T. Otsu, B. Yamada, and M. Imoto, *ibid.,* **1,** 61 (1966).

(75) K. C. Ramey and J. Messick, *J. Polymer Sci.,* **A2-4,** 1 (1966).

(76) K. Hatada, K. Ota, and H. Yuki, *J. Polymer Sci.,* **B5,** 225 (1967).

(77) K. C. Ramey, D. C. Lini, and G. L. Statton, *J. Polymer Sci.,* **A1-5,** 257 (1967).

(78) R. Yamadera and M. Murano, *J. Polymer Sci.,* **B3,** 821 (1965).

(79) K. Matsuzaki, T. Uryu, K. Ishigure, and M. Takeuchi, *J. Polymer Sci.,* **B3,** 835 (1965).

(80) K. Matsuzaki, T. Uryu, K. Ishigure, and M. Takeuchi, *J. Polymer Sci.,* **B4,** 93 (1966).

(81) J. Bargon, *Kolloid-Z.,* **213,** 51 (1966).

(82) R. Yamadera and M. Murano, *J. Polymer Sci.,* **A1,** 1059 (1967).

(83) M. Murano and R. Yamadera, *J. Polymer Sci.,* **B5,** 333 (1967).

(84) H. Sobue, T. Uryu, K. Matsuzaki, and Y. Tabata, *J. Polymer Sci.,* **B1,** 409 (1963).

(85) G. Geuskens, J. C. Lubikulu, and C. David, *Polymer,* **7,** 63 (1966).

(86) J. Brandrup and M. Goodman, *J. Polymer Sci.,* **B2,** 123 (1964).

(87) M. Goodman and J. Brandrup, *J. Polymer Sci.,* **B3,** 127 (1965).

(88) E. G. Brame, Jr., R. S. Sudol, and O. Vogl, *J. Polymer Sci.*, **A2**, 5337 (1964).

(89) H. L. Frisch, C. L. Mallows, and F. A. Bovey, *J. Chem. Phys.*, **45**, 1565 (1966).

(90) R. C. Ferguson and W. D. Phillips, *Science*, **157**, 257 (1967).

(91) T. Yoshino, M. Shinomiya, and J. Komiyama, *J. Am. Chem. Soc.*, **87**, 387 (1965).

(92) P. J. Flory and J. D. Baldeschwieler, *J. Am. Chem. Soc.*, **88**, 2873 (1966).

(93) J. W. L. Fordham, P. H. Burleigh, and C. L. Sturm, *J. Polymer Sci.*, **20**, 251 (1956).

(94) F. A. Bovey, G. V. D. Tiers, and G. Filipovich, *J. Polymer Sci.*, **38**, 73 (1959).

(95) A. Odajima, *J. Phys. Soc. Japan*, **14**, 777 (1959).

(96) T. Yoshino, H. Kyogoku, J. Komiyama, and Y. Manabe, *J. Chem. Phys.*, **38**, 1026 (1963).

(97) R. C. Ferguson, private communication.

CHAPTER II

Configurational Sequences and The Mechanism of Vinyl Propagation

1. *Sequence Statistics.* The distribution of configurational sequences in vinyl polymer chains is of considerable interest theoretically and of great significance practically, particularly since the development of Zeigler-Natta catalyst systems by means of which stereospecific polymerization can be achieved. Earlier studies[1-3] were hampered by the lack of any direct experimental measurements of configurational sequences, for usually (but not always) the polymer will crystallize and thus be amenable to X-ray study only if highly stereoregular. For less stereoregular chains it was necessary to have recourse to measurements of solubility properties,[4,5] melting point,[2,6,7] swelling behavior,[8] infrared spectra,[9-14] and chemical reactivity.[13] None of these methods lends itself to quantitative measurement, and they give no information concerning longer sequences. The application of high-resolution NMR spectroscopy to vinyl polymers[15,16] made the quantitative measurement of sequence distribution feasible, as we have seen in Chapter I, and allowed one to compare actual polymer chains with theoretical predictions and to correlate stereoregularity with other physical properties.[17-22]

The sequence statistics most commonly considered are (*a*) the Bernoulli-trial and (*b*) the first-order Markov. Perhaps it is generally understood what each of these types of sequences implies concerning the mechanism producing it, but neverthe-less it may be well to consider this question briefly here, be-cause one sometimes finds statements in the literature that are inconsistent or misleading. In Fig. II-1 is shown at the top

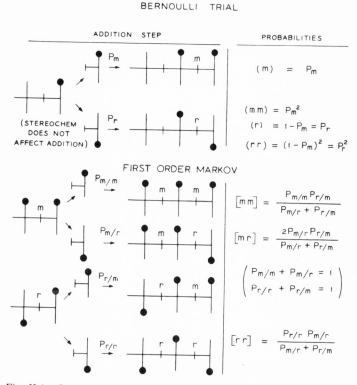

Fig. II-1. Bernoulli-trial and first-order Markov propagation steps and probabilities.

the building up of a chain by Bernoulli-trial steps. The chain end is not represented as having any particular stereochem-istry, *i.e.,* not only is it not specified as *d* or *l* (a terminology to be avoided; see p. 10) but it is also unimportant whether it is *m* or *r*. The process is thus like reaching *at random* into a large jar containing balls marked m or r; the proportion of m's in this jar is called here P_m, as defined in Chapter I. It is often said of this mechanism that the addition "is influenced only by the end-unit of the growing chain"; such statements should not be understood to mean that addition is influenced by the *stereochemistry* of the end of the growing chain, for it is not: *one* monomer chain unit considered alone has no stereochemistry, but can of course exert a steric influence. Free-radical polymerizations generally follow these simple statistics, as we have seen, although Coleman *et al.*[23] have obtained evidence for deviations in methacrylate polymerizations at low temperature, and Tsuruta *et al.*[24] have found similar deviations for methacrylates with bulky sidechains.

The first-order Markov sequence[22,23,25-29] is generated by propagation steps in which the adding monomer *is* influenced by the stereochemistry of the growing chain end, which may be m or r. We now have four probabilities characterizing the addition process: $P_{m/m}$, $P_{r/m}$, $P_{m/r}$, and $P_{r/r}$. (The designation $P_{r/m}$ means here the probability that the monomer adds in m-fashion to an r chain end, etc.) For the triad frequencies we have

$$(mm) = \frac{P_{m/m} P_{r/m}}{P_{m/r} + P_{r/m}} \tag{II-1}$$

$$(mr) = \frac{2 P_{m/r} P_{r/m}}{P_{m/r} + P_{r/m}} \tag{II-2}$$

$$(rr) = \frac{P_{r/r} P_{m/r}}{P_{m/r} + P_{r/m}} \tag{II-3}$$

We also have the relationships

$$P_{m/m} + P_{m/r} = 1 \tag{II-4}$$

and

$$P_{r/r} + P_{r/m} = 1 \tag{II-5}$$

expressing the fact that every step must go either m or r. There are thus really only two independent probabilities, which we shall take as $P_{r/m}$ and $P_{m/r}$. The propagation may be characterized if we can determine values of these conditional probabilities. One may also of course easily imagine second-order Markov processes, characterized by four conditional probabilities, and implying an influence of an additional chain unit—and still higher-order Markov propagation steps. One may in addition imagine non-Markov processes, one of which we shall discuss later.

Concerning the fitting of data to postulated mechanisms, one can say the following:

1. From *dyad* information alone, *i.e.,* (*m*) and (*r*), any mechanism can be *fitted* but none can be *tested*. By "tested," we mean that a polymer can be shown to be consistent or inconsistent with a given model, at a given level of sequence discrimination. But there is always the possibility that examination with higher discrimination may reveal inconsistencies with the proposed model.

2. From *triad* information (which implies dyad as well),*i.e.,* (*mm*), (*mr*), and (*rr*), a Bernoulli model can be *tested* (Chapter I) and Markov models of any order can be *fitted*. In addition, certain non-Markov models can be *tested*, as we shall see.

3. From *tetrad* information, a first-order Markov model can be *tested*, and higher orders fitted.

4. From *pentad* information, a second-order Markov model can be *tested*, and higher orders fitted.

The extension of such statements to higher sequences is obvious.

We have seen that from triad data, consistency with Bernoulli statistics may be tested by the relationships

$$(mm) = P_m^2 = (m)^2 \tag{II-6}$$

$$(mr) = 2P_m(1 - P_m) = 2(m)[1 - (m)] \tag{II-7}$$

$$(rr) = (1 - P_m)^2 = [1 - (m)]^2 \tag{II-8}$$

This is equivalent to fitting to the now familiar parabolic curves describing P_m as functions of triad frequency (Chapter I); dyad data are useful for support, but are not necessary.

There is an important point in connection with such fitting to Bernoullian schemes that deserves emphasis.[27] So far as passing the triad test is concerned, it is sufficient that the placement of the adding monomer unit be independent of whether the stereochemistry of the chain *end* is *m* or *r*; it can be influenced to an arbitrarily high degree by the stereochemistry of the next placement down, *i.e.,* can be subject to a very strong *antepenultimate* effect, and still appear to be Bernoullian. Thus, for example, we may imagine that by a third-order Markov process, we may generate a chain with an *mmrr* repeating unit:

$$\ldots mmrrmmrrmmrr \ldots$$

By observation of only dyad and triad frequencies, such a polymer would appear to be configurationally random or "atactic." From the observation of tetrad frequencies, however, it would become obvious that this is not the case, since from Fig. I-7 and Table I-1 we find that for a random polymer

$$(mmm) = (rrr) = (rmr) = (mrm) = \tfrac{1}{8}$$

$$(mmr) = (rrm) = \tfrac{1}{4}$$

while for the above polymer:

$$(mmr) = (rrm) = \tfrac{1}{2}$$

the other tetrad sequences being absent.

In Fig. II-2 is shown the spectrum of a polymethyl methacrylate prepared with a free-radical initiator at the high temperature of 135°. We may expect (see Table I-3) that P_m will be in excess of 0.27, and that the configurational statistics will be Bernoullian. We might first test for Bernoullian fitting by placing the (*mm*), (*mr*), and (*rr*) intensities on the curves of

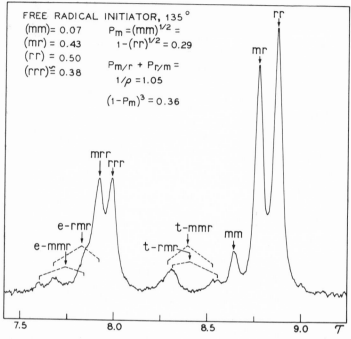

Fig. II-2. Polymethyl methacrylate prepared with a free-radical initiator at 135° (spectrum taken on 15% solution in chlorobenzene).

Fig. I-6. We may also do this in a slightly different way by calculating the conditional probabilities $P_{m/r}$ and $P_{r/m}$ from the triad data using the relationships [readily derived from eqs. (II-1) through (II-5)]:

$$P_{m/r} = \frac{(mr)}{2(mm) + (mr)} \tag{II-9}$$

$$P_{r/m} = \frac{(mr)}{2(rr) + (mr)}. \tag{II-10}$$

It is readily shown that for Bernoullian propagation, the sum of $P_{m/r}$ and $P_{r/m}$, which we shall designate as Σ_p, is equal to unity. In the present case we find

$$P_{m/r} = \frac{0.43}{2(0.07) + 0.43} = 0.75$$

$$P_{r/m} = \frac{0.43}{2(0.50) + 0.43} = 0.30$$

$$\Sigma_p = \overline{1.05}$$

This appears to be consistent with Bernoullian statistics within experimental error. We may therefore calculate

$$P_m = (mm)^{1/2} = 1 - (rr)^{1/2} = 0.29$$

We also should test the tetrad intensities for consistency with Bernoullian statistics. For this purpose, the *rrr* peak in Fig. II-2 is probably best, although none is really satisfactory (Fig. I-9*b* is much more suitable, but the integral spectrum was not run). It is found that (see Table I-1)

$$(rrr) = (1 - P_m)^3 = 0.36,$$

whereas the observed intensity of *rrr* is 0.38 ± 0.02; the agreement is as good as can be expected. Even passing the tetrad test does not rule out mechanisms of higher order than first-order Markov, as we have seen, but one usually accepts agreement at this point as indicating Bernoullian statistics, as it is generally considered undesirable to make such interpretations unnecessarily complicated. In a strict sense, however, one can *never* rule out mechanisms of a higher order than one is able to test.

In Fig. II-3 is shown the spectrum of a polymer of methyl cis-β-d_1-methacrylate:

$$\underset{CH_3O_2C}{\overset{CH_3}{\diagdown}} C = C \underset{D}{\overset{H}{\diagup}}$$

made in toluene solution at $-78°$ with phenylmagnesium bromide initiator in the presence of diethyl ether (9:1 mole

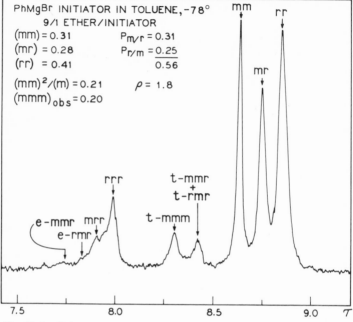

Fig. II-3. Polymethyl *cis-β-d₁*-methacrylate prepared with phenylmagnesium bromide initiator in toluene at $-78°$; 9:1 molar ratio of diethyl ether to initiator (spectrum taken on 15% solution in chlorobenzene).

ratio of ether to initiator). In this polymer, there is no observable geminal coupling, and so even the *meso* β-methylene proton resonances are singlets, somewhat broadened by unresolved D-H coupling. The interpretation of the methylene spectrum will be discussed later in this chapter. For the present, we are concerned with the triad peaks. We see at once that the *mr* peak is smaller than either the *mm* or *rr* peaks, an indication of non-Bernoullian statistics. When the triad intensities, indicated in Fig. II-3, are used to calculate the first-order Markov probabilities from eqs. (II-9) and (II-10), we obtain

$$P_{m/r} = 0.31$$
$$P_{r/m} = 0.25$$

Σ_p is 0.56 and clearly indicates a deviation from Bernoullian statistics. To test consistency with a first-order Markov mechanism, we must use tetrad intensities and the following relations:[27,28]

$$(mmm) = \frac{(1 - P_{m/r})^2 P_{r/m}}{P_{m/r} + P_{r/m}} \left[= \frac{(mm)^2}{(m)} \right] \qquad \text{(II-11)}$$

$$(mmr) = \frac{2 P_{m/r}(1 - P_{m/r}) P_{r/m}}{P_{m/r} + P_{r/m}} \left[= \frac{(mm)(mr)}{(m)} \right] \qquad \text{(II-12)}$$

$$(mrm) = \frac{P_{m/r} P_{r/m}^2}{P_{m/r} + P_{r/m}} \left[= \frac{(mr)^2}{4(r)} \right] \qquad \text{(II-13)}$$

$$(mrr) = \frac{2 P_{m/r} P_{r/m}(1 - P_{r/m})}{P_{m/r} + P_{r/m}} \left[= \frac{(mr)(rr)}{(m)} \right] \qquad \text{(II-14)}$$

$$(rmr) = \frac{P_{m/r}^2 P_{r/m}}{P_{m/r} + P_{r/m}} \left[= \frac{(mr)^2}{4(m)} \right] \qquad \text{(II-15)}$$

$$(rrr) = \frac{P_{m/r}(1 - P_{r/m})^2}{P_{m/r} + P_{r/m}} \left[= \frac{(rr)^2}{(r)} \right] \qquad \text{(II-16)}$$

It is desirable (and sufficient) to show that *four* (not five) of the left-hand relationships hold, but this is not always experimentally possible. The relationships in brackets hold for Bernoullian statistics also, and hence do not constitute a test for first-order Markov; they are useful in testing non-Markovian chains, as we shall see shortly. In the present case, (mmm) calculated from eq. (II-11) is 0.21, and the observed relative intensity is 0.20. This polymer can thus be described by first-order Markov statistics within experimental error.

Another useful quantity is the *persistence ratio*, defined by Coleman and Fox[20,21] as

$$\rho = \frac{2[m][r]}{[mr]}.$$

This is equal to $1/\Sigma_p$ for first-order Markov statistics. For this polymer, then

$$\rho = 1/0.56 = 1.8$$

This quantity is unity for a Bernoullian polymer. In the present case, both the r and m sequences persist longer than for a Bernoullian polymer, *i.e.,* the polymer tends toward the "block" type, a now well-recognized species.

It might be appropriate to summarize the four possible types of chains that can be described in first-order Markov terms; polymers resembling the last two can of course be produced by Bernoullian propagation, the limits being the corresponding pure polymer types:

1. "Block":	$P_{m/r} < 0.5$
	$P_{r/m} < 0.5$
	$\Sigma_p < 1$
2. Heterotactic-like:	$P_{m/r} > 0.5$
	$P_{r/m} > 0.5$
	$\Sigma_p > 1$
3. Isotactic-like:	$P_{m/r} < 0.5$
	$P_{r/m} > 0.5$
4. Syndiotactic-like:	$P_{m/r} > 0.5$
	$P_{r/m} < 0.5$

2. *The Coleman-Fox Propagation Mechanism.* Mechanistically, one can certainly plausibly account for the tendency of the *meso* stereochemistry to propagate itself, for one may easily imagine, as many authors have suggested, that in Grignard- or in metal-alkyl-initiated polymerizations the counter-ion chelates with the incoming monomer and the chain-end in such a way as to favor m placements. It is not quite so easy to see why the r placement would tend to propagate itself, although again complexing can be imagined which would account for this. Coleman and Fox[20,29] have proposed an alternative mechanistic scheme which is physically appealing. They suggest that in polymerizing systems (usually initiated by metal alkyls) in which runs of m's and r's are produced, *i.e.,* "block" chains, there may be two (or possibly more) states of the propagating chain end, corresponding to chelation by counter-ion and the interruption of this chelation by solvation. The intervals between arrival and departure of the solvating species (usually an added ether) are imagined to be longer than that corresponding to an average

propagation step, but not necessarily very much longer, since block lengths are often very short. To describe the statistical consequences of this mechanism, we may write modified forms of the relationships holding for simple Markov propagation.[27,28] In the two-state model, triad and tetrad proportions are as shown in Table II-1, in which the sequences are

TABLE II-1. Sequence Proportions for Coleman-Fox Two-state Model

Triads:

$$(mm) = (m)^2 + ax \qquad \text{(II-17)}$$
$$(mr) = 2(m)(r) - 2ax \qquad \text{(II-18)}$$
$$(rr) = (r)^2 + ax \qquad \text{(II-19)}$$

Tetrads:

$$(mmm) = \frac{(mm)^2 + abx^2}{(m)} \qquad \text{(II-20)}$$

$$(mmr) = \frac{(mm)(mr) - 2abx^2}{(m)} \qquad \text{(II-21)}$$

$$(mrm) = \frac{(mr)^2 + 4acx^2}{4(r)} \qquad \text{(II-22)}$$

$$(mrr) = \frac{(mr)(rr) - 2acx^2}{(r)} \qquad \text{(II-23)}$$

$$(rmr) = \frac{(mr)^2 + 4abx^2}{4(m)} \qquad \text{(II-24)}$$

$$(rrr) = \frac{(rr)^2 + acx^2}{(r)} \qquad \text{(II-25)}$$

expressed in a manner which shows clearly their deviation from expectation for Bernoullian and first-order Markov propagation (p. 12). The notation is as follows:

$$\text{probability of } m \text{ placement} = (m) = \frac{(\lambda_a k_{1i} + \lambda_b k_{2i})}{(\lambda_a k_1 + \lambda_b k_2)}$$

where λ_a = rate constant for State 2 $\rightarrow$ State 1 of the chain end

λ_b = rate constant for State 1 $\rightarrow$ State 2 of the chain end

k_{1i} = rate constant for m placement in State 1

k_{2i} = rate constant for m placement in State 2

k_1 = $k_{1i} + k_{1s}$, where k_{1s} is the rate constant for r placement in State 1

k_2 = $k_{2i} + k_{2s}$, where k_{2s} is the rate constant for r placement in State 2

$$a = \frac{\lambda_a \lambda_b k_1 k_2}{(\lambda_a k_1 + \lambda_b k_2)^2} \left(\frac{k_{1i}}{k_1} - \frac{k_{2i}}{k_2} \right)^2$$

$b = k_{1i} k_{2i} / k_1 k_2$

$c = k_{1s} k_{2s} / k_1 k_2$

$d = (\lambda_a k_1 + \lambda_b k_2)/k_1 k_2$

$x = [M]/([M] + d)$, where $[M]$ is the monomer concentration in the polymerizing system

It is assumed that the polymerization takes place in a homogeneous fashion, which is probably reasonable for the alkyllithium-initiated polymers to which the treatment has been applied. Coleman and Fox[20] outline how the rate constants could be determined individually, assuming that λ_a and λ_b, but not the k's, may depend on the concentration of the initiator.

Such a polymerizing system will *in general* generate chains which cannot be described by first-order Markov statistics or by Markov statistics of any finite order. Let us, however, imagine the particular case (which is a plausible one) that when the chain end is in State 1 only one type of propagation —say isotactic—can occur, while in State 2 only syndiotactic propagation can occur. Under these conditions, b and c become zero, but a and x are still nonzero and positive. The extra terms in the tetrad relationships [eqs. (II-20) to (II-25), Table II-1] become zero, so these revert to the bracketed relationships in eqs. (II-11) to (II-16), *i.e.,* to first-order Markov statistics. The extra terms in the dyad-triad relationships [eqs. (II-17) to (II-19)] remain, but the corresponding deviations from Bernoullian statistics cannot be told from those that would be expected for first-order Markov statistics. In other words, the resulting chain has a structure indistinguishable from that which would be generated on the assumption of a "penultimate" effect, but without assuming such an effect. It therefore follows that adherence to first-order Markov statistics does not in itself enable one to decide whether the polymerization occurs by a two-state mechanism or by a one-state mechanism with influence of the chain configuration on the monomer placement.

When a approaches zero, *i.e.*, when the arrival and departure intervals of the solvating species become short compared to the propagating steps, the polymer statistics will approach Bernoullian.

A polymer prepared with phenylmagnesium bromide in toluene at $-78°$ and apparently consistent with a two-state mechanism is represented by Figs. I-5a and I-9a for which we find $(mm) = 0.75$, $(mr) = 0.12$, $(rr) = 0.13$ and $(mmm) = 0.78$, the last being the most readily measurable of the tetrad resonances. From these data, we find

$$P_{m/r} = 0.09$$
$$P_{r/m} = 0.40$$
$$\Sigma_p = \overline{0.49}$$
$$\rho = 2.56$$
$$\bar{n} = 1/(mr) = 8.3$$
$$(mm)^2/(m) = 0.68$$

We see that (mmm) exceeds expectation for a first-order Markov process by 0.10. It follows from eq. (II-20) that $abx^2 \simeq 0.08$. If all tetrad intensities could be measured with accuracy, a, b, c, and x could be determined. It has not yet been ascertained, however, whether this polymer conforms to second-order Markov statistics. This can be tested using pentad frequencies, as we have seen (p. 44), but this has not yet been done.*

*NOTE ADDED IN PROOF: Since the above paragraph was written it has been shown [by H. L. Frisch, C. L. Mallows, F. Heatley, and F. A. Bovey, *Macromolecules*, **1**, 533 (1968)] that this polymer does not quite conform to second-order Markov statistics, but does conform very precisely to a Coleman-Fox mechanism with very reasonable parameters.

In the preparation of this polymer (which was obtained indirectly from the Rohm and Haas laboratories) no ether was deliberately added,[30] but, as we shall see a little later in this chapter, a significant amount of ether will be retained by the Grignard reagent unless special precautions are taken. The present indications seem to be, then, that the widest deviations from first-order Markov statistics occur in the presence of small quantities of complexing solvent. When the concentration of complexing solvent is increased, syndiotactic block sequences increase in length (see Fig. II-3 and related discussion; also ref. 23) but the chain obeys first-order Markov statistics. This may be because two-site propagation still occurs, but b and c have become zero, *i.e.,* each site permits only m (or r) propagation, as previously discussed.

Again, Coleman et al.[23] have observed that a polymethyl methacrylate prepared with 9-fluorenyllithium in 95:5 toluene–tetrahydrofuran (by volume) at $-60°$ showed a significant deviation from first-order Markov statistics. They defined the quantity Ω as a measure of this deviation:

$$\Omega = \frac{(r)(rrr)}{(rr)^2}$$

and found it to be 1.09; when the tetrahydrofuran concentration was increased to 10%, Ω decreased to unity within experimental error, corresponding to first-order Markov (or Bernoulli) propagation, from eq. (II-16).

3. *The Direction of Addition to the Double Bond; β-Carbon Stereochemistry.* It was first demonstrated in 1958 by Fox and co-workers[31-33] that the anionic polymerization of methyl methacrylate at low temperature using metal-alkyl initiators can give various crystallizable forms of the polymer, depending upon the composition of the polymerization solvent. Polymer prepared in the strongly solvating solvent ethylene glycol dimethyl ether was highly syndiotactic. Polymer prepared in toluene was highly isotactic. Polymer prepared in toluene containing small proportions of ethers contained both isotactic and syndiotactic blocks, such as we have already

discussed. These observations have been extended to a number of acrylic and methacrylic monomers[34-37] and the mechanism of the reaction has received extensive study.[15,16,38-53] It is clear from this work that the polymerization mechanism is complex. Rapid reaction of the metal alkyl and monomer occurs, but in addition to propagation to produce long chains, several side reactions may occur, including reaction at carbonyl as well as olefinic double bonds,[38,40,43,45] leading to substantial proportions of low-molecular-weight products. These side reactions can be minimized by appropriate measures,[38,40,41] and we shall not concern ourselves further with them here. Those chains which do grow to form high polymer apparently do so at a wide range of rates, for the molecular-weight distribution is very broad.[38,43,47,52]

In addition to kinetic and molecular-weight measurements, the stereochemical configuration of the chains has also been observed in many of these studies, but only with respect to the α-carbon atoms, since this of course is the usual basis for stereochemical classification. However, the β-methylene protons must also be considered for a complete understanding of the propagation pathway. At first sight, this might appear to be rather academic since the β-carbon stereochemistry can be observed only if the monomer is labelled with deuterium, and has no practical significance with regard to any properties of the polymer other than its spectroscopic properties. However, the β-carbon stereochemistry is very important with respect to the mode of addition of the growing chain end to the double bond of the monomer. Natta *et al.*[53] and Peraldo and Farina[54] observed that the polarized infrared spectra of polymers of *cis*-1-d_1-propylene and *trans*-1-d_1-propylene are quite different. Miyazawa and Ideguchi[55] showed that *cis*-1-d_1-propylene yields the *erythro*diisotactic polymer

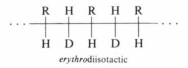

*erythro*diisotactic

and that *trans*-1-d_1-propylene yields the *threo*diisotactic polymer

*threo*diisotactic

These results were interpreted as indicating that, with the Ziegler-Natta catalyst employed, *cis* opening of the double bond occurs. The same conclusion was reached by Natta *et al.*[56] for the polymerization of *trans*-propenyl isobutyl ether and *cis*- and *trans*-β-chlorovinyl alkyl ethers.[57] It is very easy to see that, in terms of simple "ball-and-stick" concepts, the steric relationships shown below must hold for *meso* β-methylene units:[57]

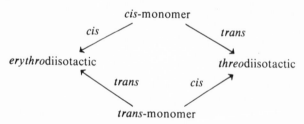

For racemic placements, there is no distinction between *erythro* and *threo* protons, at least for long stereoregular chains, and so only one syndiotactic polymer is possible, regardless of the mode of addition. In the *mrr* tetrad, a distinction is possible in principle, and may actually be observable at 220 MHz; see H. L. Frisch, C. L. Mallows, F. Heatley, and F. A. Bovey, *Macromolecules,* **1,** 533 (1968).

Such studies have now been extended to acrylate and methacrylate monomers.[58–64] In the work to be described here,[58,59] the following three monomers have been employed, all being *cis*:

II.
$$\begin{array}{c} CH_3 \\ \\ CH_3OOC \end{array} \!\! C\!=\!C \!\! \begin{array}{c} H \\ \\ D \end{array}$$

III.
$$\begin{array}{c} CH_3 \\ \\ CH_3CH_2OOC \end{array} \!\! C\!=\!C \!\! \begin{array}{c} H \\ \\ D \end{array}$$

Work was initiated with the isopropyl dideuteroacrylate (**I**),[58] and more recently[59] has dealt mainly with the methacrylates, of which **III** has been chiefly employed because of greater ease of synthesis. In the polymer NMR spectra, principal attention centers on the β-proton spectrum, with the side-chain resonances providing important supporting information concerning triad sequence frequencies. Spectra were mostly obtained at 60 MHz. A few were run at 100 MHz. Pentad intensities were therefore not distinguishable. As we have seen in Chapter I (p. 19), the β-proton in the *erythro*diisotactic configuration appears at lower field than in the *threo*diisotactic configuration in polyacrylate spectra, and it has been assumed that this is true for methacrylate chains as well. The β-protons at lower field will themselves be referred to as "*erythro*" (abbreviated e) and the more shielded β-protons will be referred to as "*threo*" (abbreviated t), as already described.

In Fig. II-4 are shown 60 MHz spectra of polyisopropyl-α-*cis*-β-dideuteroacrylate (*b*, *c*, and *d*) together with the observed spectrum (*a*) of the non-deuterated polymer, taken from Fig. I-16. The singlets at 7.86τ and 8.32τ in (*d*) and (*f*) correspond to e and t protons. The polymer is highly isotactic (see Chapter I, Sec. 6), so these are e-*mmm* and t-*mmm*, respectively. There is no clearly distinguishable r β-methylene resonance (probably mainly *mrm*), which would be expected from other studies to appear at *ca.* 8.20τ. Whether this resonance might be shifted from this position in a predominantly isotactic chain[65] has not yet been established experimentally, but it does not appear that this polymer is likely to be less than *ca.* 95% isotactic. Spectrum (*d*) is the calculated spectrum for the normally expected deuterium

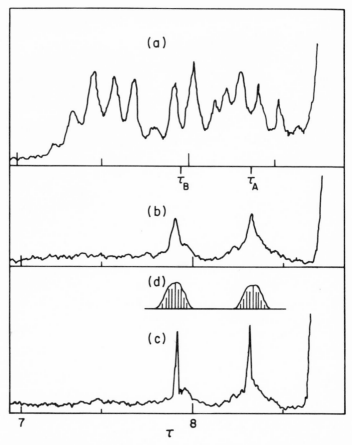

Fig. II-4. Spectra of isotactic poly-*iso*propyl-α,β-*cis*-d_2-acrylate (14% solution in chlorobenzene at 150°). (*a*) Nondeuterated polymer; (*b*) deuterated polymer, not decoupled; (*c*) deuterated polymer with irradiation of deuterium; (*d*) spectrum calculated from D-H couplings.

couplings. It can be seen that quadrupolar relaxation of the deuterium nucleus, particularly effective in slow-moving polymer chains, has partially decoupled the deuterium nuclei from the protons; in spectrum (*c*), decoupling has been made complete by irradiation of the deuterium nuclei at 9.1 MHz. The lines are now sharp.

From the mechanistic viewpoint, the surprising thing about this spectrum is the nearly equal intensity of the peaks. One might expect addition to be either *cis* or *trans*, but not both. It was found that the mode of addition is strongly dependent upon the concentration of diethyl ether. Although the Grignard reagent was dried in vacuum, a small amount of ether remained in the polymerizing system producing the polymer shown in Fig. II-4. When this was rigorously removed by baking the initiator in vacuum, there was produced a polymer giving the spectrum shown in Fig. II-5 (*c*). [Spectrum (*b*) is the same as in Fig. II-4]. Now, *cis* addition is strongly preferred. When ether is added in 10:1 mole ratio to the Grignard, the polymer produced shows a very strong *threo*

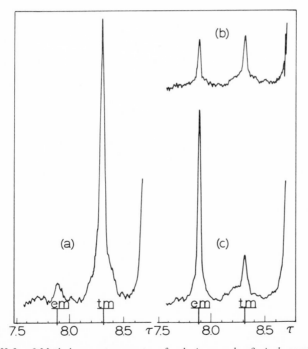

Fig. II-5. β-Methylene proton spectra of poly-*iso*propyl-α,β-*cis*-d_2-acrylate prepared with phenylmagnesium bromide and (*a*) large molar excess of diethyl ether; (*b*) partial ether removal; and (*c*) rigorous ether removal.

peak [Fig. II-5, spectrum (*a*)] and has been produced almost exclusively by *trans* addition. All polymers are isotactic. *Thus, by varying the ether content in the system, one can control the β-configuration without affecting the α-configuration.* The Grignard-initiated polymerization system is probably heterogeneous, at least in the absence of ethers or at low ether levels, and is undoubtedly very complex. Yoshino *et al.*[63] have found that the polymerization temperature has a profound effect on the ratio of *cis* to *trans* addition. In toluene (no ether present), *cis* addition is the exclusive pathway at −83°, in agreement with the result shown in Fig. II-5*c*. Above about −60°, *trans* addition is preferred, while near −60° a polymer giving a spectrum like Fig. II-5*b* was obtained. It was furthermore found that it is the temperature within the initial 1 min. of polymerization which determines the subsequent mode of propagation, for the mode determined during that phase of the reaction will persist even if the temperature is changed after 1 min. This behavior was attributed[64] to the fact that the fraction of surviving Ph-Mg bonds is greater at low temperatures, these being evidently responsible for the *cis* double bond opening. The effect of higher temperature is thus parallel to that of ether, but the detailed reasons for this parallelism remain to be elucidated.

For polymers such as shown in (*d*) and (*f*) in Fig. II-4 in which both *erythro* and *threo* methylenes are clearly present, an interesting question arises: are these sequences distributed at random or do they occur in long runs, perhaps even as entire *erythro* and entire *threo* chains? To study this question, Yoshino and Kuno[62] have carried out a most ingenious experiment. Using isopropyl *trans*-β-monodeutero acrylate— i.e., still having an α-proton—it was shown from the coupling patterns in the polymer spectra that virtually all the *threo* units have *threo* units as neighbors and all the *erythro* units have *erythro* units as neighbors. Thus, they must be produced in long runs. Perhaps there are certain catalyst sites that turn out only *threo* chains and others that turn out only *erythro* chains. Or perhaps a given site changes its properties periodically. If this is the case, it must do so at a rate that is much lower than the rate of propagation. This raises the possibility of two interleaved Coleman-Fox mechanisms[20,29] operating separately at the α- and β-carbon atoms.

Because of the complexities of the Grignard reagent, most of the work of Schuerch *et al.*[58,59] was carried out with metal-alkyl initiators. Using 9-fluorenyllithium related but apparently opposite effects are observed. Preliminary work with isopropyl acrylate showed that increasing proportions of tetrahydrofuran in the polymerization solvent caused *cis* addition rather than *trans* to be increasingly preferred. The origin of this effect is probably quite different from that of ether in the Grignard-initiated polymerization.

More extensive studies of solvent effects on α- and β-carbon configuration have been carried out with the *cis*-β-d_1 methacrylates[59] (II and III, p. 57). The data are summarized in Table II-2. The polymerization of III in toluene at $-78°$ with 9-fluorenyllithium initiator produces a strongly isotactic polymer (*a* in Table II-2), (*m*) being 0.87 (Fig. II-6*a*). The *m* protons are 88% *threo*, indicating that *trans* addition is very strongly preferred. In this "living" polymer system,[66,67] viscosity,[59] spectroscopic,[68] and conductance[59,69] measurements indicate that contact ion pairs, not specifically solvated, are the chain carriers:

$$C_6H_5 \left[CHD-\underset{\underset{CO_2R}{|}}{\overset{\overset{CH_3}{|}}{C}} \right]_n CHD-\underset{\underset{CO_2R}{|}}{\overset{\overset{CH_3}{|}}{C^+}} Li^-$$

There is no complexing of the living chains to form dimers and higher aggregates as with living hydrocarbon polymer chains.[70,71] Under these conditions a single species is responsible for chain growth, and the chain statistics should be Bernoullian or first-order Markov. Triad analysis indicates the former, with $P_m = 0.87$, and there is thus no evidence of "penultimate" effects. (The α-methyl spectrum, not shown in Fig. II-6, is complicated by the overlapping ester methyl triplet, but can still be analyzed.) In the *r*-tetrad region, the only discernible peak is *mrm*; thus the flaws in the stereo-

TABLE II-2. Anionic Polymerization of Ethyl *cis-β-d₁*-Methacrylate (ref. 59)

	Initiator	Initiator conc., M	THF, mole ratio to initiator	Solvent	Temp., °C	$\dfrac{(tm)†}{(m)}$	(m)	(mm)	(mr)	(rr)	Configuration statistics*
a	Fl-Li	0.00074	0	toluene	−78°	0.88	0.87	0.79	0.16	0.04	P_m: 0.87
b	Fl-Li	0.047	7.5	toluene	−78°	0.11	0.64	0.55	0.18	0.27	Σ_p: 0.39
c	Fl-Li	0.047	7.5	toluene	+30°	0.09	0.80	0.72	0.17	0.11	Σ_p: 0.58
d	Fl-Li	0.00055	2.2×10^4	THF	−78°	0.51	0.15	0.05	0.21	0.74	P_m': 0.15
e	Fl-Cs	0.021	520	THF	−78°	0.70	0.28	<0.05	0.48	0.48	—

*If polymer is Bernoullian, P_m is given; if polymer is first-order Markov (or nearly so), Σ_p ($= P_{m/r} + P_{r/m}$) is given (see p. 46).

†$tm = threo\text{-}meso$.

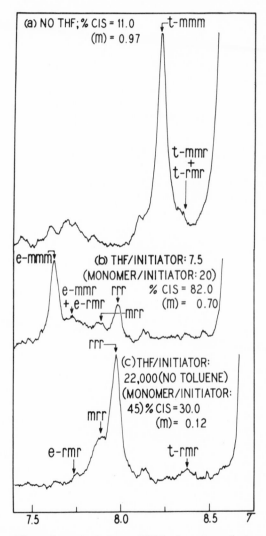

Fig. II-6. Effect of tetrahydrofuran (THF) on polymerization of ethyl *cis-d*$_1$-methacrylate in toluene at $-78°$ with 9-fluorenyllithium; (*a*) no THF; (*b*) THF : initiator ratio 7.5; (*c*) THF : initiator ratio 2.2×10^4 (no toluene).

regularity are of the type

$$\cdots \begin{array}{c} CH_3 \\ | \\ | \\ CO_2R \end{array} \begin{array}{c} H \\ \\ D \end{array} \begin{array}{c} CH_3 \\ | \\ | \\ CO_2R \end{array} \begin{array}{c} H(D) \\ \\ D(H) \end{array} \begin{array}{c} CO_2R \\ | \\ | \\ CH_3 \end{array} \begin{array}{c} D \\ \\ H \end{array} \begin{array}{c} CO_2R \\ | \\ | \\ CH_3 \end{array} \cdots$$

and they occur on the average about once every ten monomer placements.

In the presence of tetrahydrofuran at a 7.5 mole ratio to the 9-fluorenyllithium (the latter being 0.047 M), the polymer remains predominantly isotactic [$(m) = 0.64$; see Fig. II-6b] but the m protons are now 89% *erythro* (b in Table II-2). The polymer departs markedly from Bernoullian ($P_{m/r} = 0.14$; $P_{r/m} = 0.25$; $\Sigma_p = 0.39$). A two-state propagation appears to be the most logical explanation. We shall consider in a moment what these two states may be. Note also that when the same polymerization is carried out at $+30°$ (c in Table II-2) instead of $-78°$, the result is much the same except that the m stereoblocks are somewhat longer.

When the polymerization is carried out in tetrahydrofuran (mole ratio to 9-fluorenyllithium: 22,000) at $-78°$, the polymer is highly syndiotactic [$(m) = 0.15$; see Fig. II-6c; d in Table II-2]; the few *meso* placements appear to occur by *cis* and *trans* addition with about equal probability, as in free-radical propagation.[60] The configurational sequence is Bernoullian. In tetrahydrofuran, as in toluene, the propagating species show negligible conductance[59,69] and must likewise be ion pairs in which the ions are solvated and separated by the solvent but are not free to move independently.[68,69] Propagation by this species exhibits no steric control by the counter-ion and no "penultimate" effect; both r placement and m placement probably proceed by both *cis* and *trans* addition; the spectrum, however, is not informative with regard to r placements.

The most interesting questions are raised by the intermediate system (b and c), in which a relatively small concentration of tetrahydrofuran is present. The simplest conclusion would appear to be that both contact ion pairs and solvated ion pairs are present, these species interchanging in the manner contemplated in the Coleman-Fox mechanism (previous section). This is too simple an explanation, however, for if this were the case the *meso* protons would be mostly *threo*, whereas in fact they are mostly *erythro*. The mechanism is thus more complex than can be explained by two sites or two species, and it is necessary to invoke a third species (for which there is as yet no spectroscopic evidence[68]), which we shall term a *peripherally solvated* ion pair. This species propagates by *m* placements, as does the contact ion pair, but does so by *cis* addition. It appears that this species may exist in lesser proportion at −78° than at 30°, being partially converted to solvent separated ion pairs when the temperature is lowered, as indicated by an increase in (*rr*).

4. *The Significance of Cis and Trans Addition and the Nature of the Propagation Species.* It should be realized that the terms *cis* and *trans* do not have the same precise meanings in terms of transition-state structures when describing vinyl monomer propagation as they have for a simple four-center reaction. It is helpful to distinguish between the *true* mode of addition, *i.e.*, the actual direction in space along which the active chain-end and monomer approach each other, and the *apparent* mode of addition, *i.e.*, the direction of addition as conventionally judged by the stereochemical configuration about the resulting bond. We shall assume that, primarily for steric reasons, the *true* mode of addition is always *trans*, as represented at the top of Fig. II-7. Then, in the formation of the next *meso* dyad, *the β-carbon configuration will depend on the mode of presentation or approach of the monomer to the growing chain-end.* (If a *racemic* dyad is formed the β-configuration of the dyad is of course indeterminate.) The growing chain-end is assumed to be effectively planar but to have a definite and fixed stereochemical relationship to the next unit (not specifically represented in Fig. II-7), *i.e.*, the end of the chain is either *m* or *r* and remains so during placement of the next monomer unit. The α-carbon configuration may of course also be determined by the monomer approach mode. The monomer might approach in a syndiotactic-like or race-

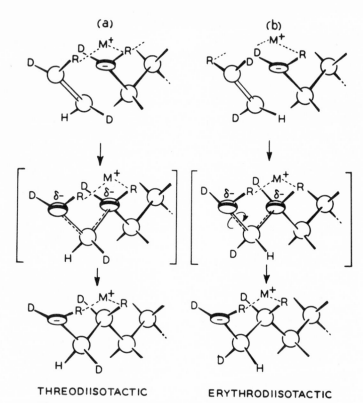

THREODIISOTACTIC   ERYTHRODIISOTACTIC

Fig. II-7. Monomer-approach modes in isotactic placement (I).

mic manner, as in (*b*), and add on in this fashion without
further change, giving an *r* dyad. But in the presence of an
effectively chelating counter-ion (represented as M^+) rotation of the newly formed chain-end could occur, giving an
erythro-meso placement. *Thus, the α-carbon configuration
depends upon chain-end rotation as well as approach mode.*
On the other hand, if the mode of approach is "isotactic-
like," as in (*a*) in Fig. II-7, the *meso* protons in the resulting
dyad will be *threo*. Thus, whether the addition is apparently
cis or *trans* depends, according to this picture, on the mode
of presentation of the monomer.

Figure II-7 is actually of course a considerably over-simplified view of the growing chain-end. A somewhat more realistic picture (but a little more difficult to visualize) is given for the contact ion pair (*a*) and the peripherally solvated ion pair (*d*) in Fig. II-8. In (*a*) the Li^+ is coordinated with the carbonyl oxygen atoms of the end and penultimate monomer units and with the carbonyl of the incoming monomer; this latter feature is represented also in a general way in Fig. II-7*a*. In hydrocarbon solvents, the counter-ion in this way not only encourages rotation of the chain-end in the event of syndiotactic-like approach, as discussed above, but also is able to guide the incoming monomer so as to make isotactic-like approach strongly preferred. In the absence of solvating molecules this guidance is *ca.* 90% effective; chain-end rotation is not necessary, and consequently the polymer is *threo*diisotactic. As solvating molecules (tetrahydrofuran) are added, the monomer-approach control function of the counter-ion begins to be seriously disturbed and eventually abolished, because in the peripherally solvated ion pair (*d*) the lithium is fully coordinated by tetrahydrofuran and the chain-end carbonyl groups, and does not coordinate with the monomer to be added. Syndiotactic-like approach, which is assumed to be inherently strongly preferred (as in free-radical propagation) becomes possible. If this happens, chain-end rotation is still fairly effective and the polymer remains predominantly isotactic, but becomes largely or almost entirely *erythro*diisotactic. Finally, at high tetrahydrofuran concentrations, all influence of the counter-ion ceases, and the polymer becomes predominantly syndiotactic. (The remaining *meso* sequences are, as we have seen, about equally *erythro* and *threo*.) In the solvent-separated ion pair which now predominates, the coordination of the Li^+ is entirely with the tetrahydrofuran; Glusker *et al.*[43] find that four tetrahydrofuran molecules are associated with each living chain.

It should be pointed out that since propagation by species (*a*) appears to be a Bernoulli-trial process, the coordination of the Li^+ with the penultimate monomer unit does not influence

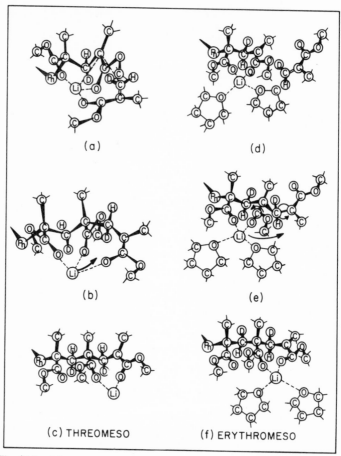

(a)

(d)

(b)

(e)

(c) THREOMESO

(f) ERYTHROMESO

Fig. II-8. Monomer-approach modes in isotactic placement (II). (a) An isotactic-like approach of the monomer to the chelated contact ion pair. (b) The new C—C bond has been formed with the methylene D on the same side of the zigzag as the ester function. (c) The Li$^+$ moves up to the new anion, with concurrent rotation of the new penultimate ester group, forming the same chelated structure as in a. (d) A syndiotactic-like approach of the monomer to the peripherally solvated contact ion pair. There is no coordination of the monomer carbonyl with the counter-ion, and nonbonded interactions force the approach into syndiotactic-like. (e) The new C—C bond has formed. (f) The Li$^+$ and its peripheral solvent shell moves up to the terminal unit, with concurrent rotation of the ester function. As the new anion resides largely on the carbonyl, there is a simultaneous rotation about the new α,β bond to reduce charge separation. This results in an *erythro-meso* placement, the methylene D being on the opposite side of the zigzag from the ester groups and the α carbon now in an incipient isotactic configuration.

its ability to coordinate with the incoming monomer or to rotate the chain-end in the event of syndiotactic-like approach, *i.e.,* it does not matter if the growing chain-end is *m* or *r*.

The chelated model of the growing chain-end which we have described here is in some ways similar to earlier proposals of Goode *et al.*,[45] Cram and Kopecky,[72] Cram,[73] and Bawn and Ledwith.[74] These models, however, do not provide an explanation for the behavior of the β-carbon configuration as a function of solvent composition.

The tendency to form solvent-separated ion pairs varies inversely as the ionic radius,[68,69] *i.e.,*

$$Li > Na > K > Cs.$$

Fluorenylcesium exists in tetrahydrofuran largely as a contact ion pair, and the propagating methacrylate polymer chain probably has a similar structure. This feature of its behavior might lead one to expect isotactic propagation even in tetrahydrofuran. The spectrum (Fig. II-9) and the sequence

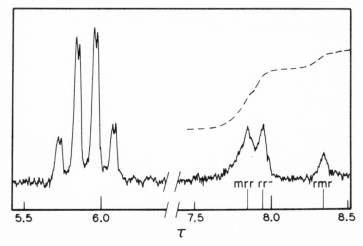

Fig. II-9. Spectrum of polyethyl methacrylate prepared with fluorenylcesium initiator in tetrahydrofuran at $-78°$. (Spectrum taken in chlorobenzene at 150°.) The broken line is the integral spectrum in the CH_2 region.

data (*e* in Table II-2) show that this is by no means the case, probably because the Cs$^+$ counter-ion also coordinates only weakly with the chain-end carbonyl groups. The sequence statistics are rather bizarre, as the polymer shows only *mr* and *rr*, *mm* being vanishingly small. (In Fig. II-9, this can be seen to be consistent with the *mr* and *rr* ester methylene quartets of approximately equal intensity at *ca.* 5.9τ.) The relative intensities of the *mrr*, *rrr*, and *rmr* tetrad peaks are approximately 2:1:1. Among the many configurational sequences which could account for this result, a structure produced by regular repetition of the unit ...*mrr*... is the simplest. The statistics of "pure" polymers of these more complex types is discussed in detail by Frisch *et al.*[27] A more exacting test would be to examine the α-methyl or ester resonances at higher fields in which pentad sequences can be discriminated (p. 14); if only *mrrr* is found, the case is proved. Such a polymer could be generated only by a second-order (or higher) Markov process.

REFERENCES FOR CHAPTER II

(1) H. L. Frisch, C. Schuerch, and M. Szwarc, *J. Polymer Sci.*, **11**, 559 (1953).

(2) B. D. Coleman, *J. Polymer Sci.*, **31**, 155 (1958).

(3) J. W. L. Fordham, *J. Polymer Sci.*, **39**, 321 (1959).

(4) S. Okamura and T. Hagashimura, *J. Polymer Sci.*, **46**, 539 (1960).

(5) H. Elias, M. Dobler, and H. Wyss, *J. Polymer Sci.*, **46**, 264 (1960).

(6) S. Newman, *J. Polymer Sci.*, **47**, 111 (1960).

(7) R. L. Miller, *J. Polymer Sci.*, **57**, 975 (1962).

(8) R. L. Miller, *Polymer*, **1**, 135 (1960).

(9) G. Natta, *Chim. e Ind. (Milan)*, **37**, 88 (1955).

(10) S. Krimm, A. R. Berens, V. L. Folt, and J. J. Shipman, *Chem. Ind. (London)*, **1958**, 512; **1959**, 433.

(11) J. W. L. Fordham, P. H. Burleigh, and C. L. Sturm, *J. Polymer Sci.*, **41**, 73 (1959).

(12) U. Baumann, H. Schreiber, and K. Tessmar, *Makromol. Chem.*, **36**, 81 (1959).

(13) G. Smets, *Angew. Chem.* (English Ed.), **1**, 306 (1962).

(14) G. Natta, G. Dall'Asta, G. Mazzanti, U. Giannini, and S. Cesca, *Angew. Chem.*, **71**, 205 (1959).

(15) F. A. Bovey and G. V. D. Tiers, *J. Polymer Sci.*, **44**, 173 (1960).

(16) F. A. Bovey and G. V. D. Tiers, *Fortschr. Hochpolymer. Forsch.*, **3**, 139 (1963).

(17) R. L. Miller and L. E. Nielsen, *J. Polymer Sci.*, **46**, 303 (1960).

(18) R. L. Miller, *J. Polymer Sci.*, **56**, 375 (1962).

(19) R. L. Miller, *SPE (Soc. Plastics Engrs.) Trans.*, **1963**, 123.

(20) B. D. Coleman and T. G. Fox, *J. Chem. Phys.*, **38**, 1065 (1963).

(21) B. D. Coleman and T. G. Fox, *J. Polymer Sci.*, **A1**, 3183 (1963).

(22) F. P. Price, *J. Chem. Phys.*, **36**, 209 (1962).

(23) B. D. Coleman, T. G. Fox, and M. Reinmöller, *J. Polymer Sci.*, **B4**, 1029 (1966).

(24) T. Tsuruta, T. Makimoto, and H. Kanai, *J. Macromol. Sci.*, **1**, 31 (1966).

(25) U. Johnsen, *Kolloid-Z.*, **178**, 161 (1961).

(26) D. Braun, M. Herner, U. Johnsen, and W. Kern, *Makromol. Chem.*, **51**, 15 (1962).

(27) H. L. Frisch, C. L. Mallows, and F. A. Bovey, *J. Chem. Phys.*, **45**, 1565 (1966).

(28) F. A. Bovey, *Pure and Applied Chem.*, **15**, 349 (1967).

(29) B. D. Coleman and T. G. Fox, *J. Am. Chem. Soc.*, **85**, 1241 (1963).

(30) D. Glusker and W. E. Goode, private communication.

(31) T. G. Fox, B. S. Garrett, W. E. Goode, S. Gratch, J. F. Kincaid, A. Spell, and J. D. Stroupe, *J. Am. Chem. Soc.*, **80**, 1768 (1958).

(32) B. S. Garrett, W. E. Goode, S. Gratch, J. F. Kincaid, C. L. Levesque, A. Spell, J. D. Stroupe, and W. H. Watanabe, *J. Am. Chem. Soc.*, **81**, 1007 (1959).

(33) J. D. Stroupe and R. E. Hughes, *J. Am. Chem. Soc.*, **80**, 2341 (1958).

(34) M. L. Miller and C. E. Rauhut, *J. Polymer Sci.*, **38**, 63 (1959).

(35) K. Butler, P. R. Thomas, and G. J. Tyler, *J. Polymer Sci.*, **48**, 357 (1960).

(36) Y. Nakayama, T. Tsuruta, J. Furukawa, A. Kawasaki, and G. Wasai, *Makromol. Chem.*, **43**, 76 (1961).

(37) D. M. Wiles and S. Bywater, *Polymer*, **3**, 175 (1962).

(38) B. J. Cottam, D. M. Wiles, and S. Bywater, *Can. J. Chem.*, **41**, 1905 (1963).

(39) D. M. Wiles and S. Bywater, *J. Phys. Chem.*, **68**, 1983 (1964).

(40) D. M. Wiles and S. Bywater, *Trans. Faraday Soc.*, **61**, 150 (1965).

(41) D. L. Glusker, E. Stiles, and B. Yoncoskie, *J. Polymer Sci.*, **49**, 297 (1961).

(42) D. L. Glusker, I. Lysloff, and E. Stiles, *ibid.*, **49**, 315 (1961).

(43) D. L. Glusker, R. A. Galluccio, and R. A. Evans, *J. Am. Chem. Soc.*, **86**, 187 (1964).

(44) W. E. Goode, F. H. Owens, R. P. Fellman, W. H. Snyder, and J. E. Moore, *J. Polymer Sci.*, **46**, 317 (1960).

(45) W. E. Goode, F. H. Owens, and W. L. Myers, *ibid.*, **47**, 75 (1960).

(46) A. Nishioka, H. Watanabe, K. Abe, and Y. Sono, *ibid.*, **48**, 241 (1960).

(47) C. F. Ryan and P. C. Fleischer, Jr., *J. Phys. Chem.*, **69**, 3384 (1965).

(48) V. D. Braun, M. Herner, V. Johnson, and W. Kern, *Makromol. Chem.,* **51,** 15 (1962).

(49) R. K. Graham, D. L. Dunkelberger, and J. R. Panchak, *J. Polymer Sci.,* **59,** 43 (1962).

(50) T. Tsurata, T. Makimoto, and Y. Nakagama, *J. Chem. Soc. Japan, Ind. Chem. Sect.,* **68,** 1113 (1965).

(51) J. Furukawa, *Polymer,* **3,** 487 (1962).

(52) T. J. R. Weakley, R. J. P. Williams, and J. D. Wilson, *J. Chem. Soc.,* **1960,** 3963.

(53) G. Natta, M. Farina, and M. Peraldo, *Atti Acad. Nazl. Lincei, Rend. Classe Sci. Fis. Mat. Nat.,* [8] **25,** 424 (1958).

(54) M. Peraldo and M. Farina, *Chim. Ind. (Milan),* **42,** 1349 (1960).

(55) T. Miyazawa and Y. Ideguchi, *J. Polymer Sci.,* **B1,** 389 (1963).

(56) G. Natta, M. Farina, and M. Peraldo, *Chim. Ind. (Milan),* **42,** 255 (1960).

(57) G. Natta, M. Peraldo, M. Farina, and G. Bressan, *Makromol. Chem.,* **55,** 139 (1962).

(58) C. Schuerch, W. Fowells, A. Yamada, F. A. Bovey, and F. P. Hood, *J. Am. Chem. Soc.,* **86,** 4481 (1964).

(59) W. Fowells, C. Schuerch, F. A. Bovey, and F. P. Hood, *J. Am. Chem. Soc.,* **89,** 1396 (1967).

(60) T. Yoshino, J. Komiyama, and M. Shinomiya, *ibid.,* **86,** 4482 (1964).

(61) T. Yoshino, M. Shinomiya, and J. Komiyama, *ibid.,* **87,** 387 (1965).

(62) T. Yoshino and K. Kuno, *ibid.,* **87,** 4404 (1965).

(63) T. Yoshino and J. Komiyama, *ibid.,* **88,** 176 (1966).

(64) T. Yoshino and J. Komiyama, *J. Polymer Sci.,* **B4,** 991 (1966).

(65) P. J. Flory and J. D. Baldeschwieler, *J. Am. Chem. Soc.,* **88,** 2873 (1966).

(66) M. Szwarc, M. Levy, and R. Milkovich, *J. Am. Chem. Soc.,* **78,** 2656 (1956).

(67) M. Szwarc, *Nature,* **178,** 1168 (1956).

(68) T. E. Hogen-Esch and J. Smid, *J. Am. Chem. Soc.,* **88,** 307 (1966).

(69) T. W. Hogen-Esch and J. Smid, *ibid.,* **88,** 318 (1966).

(70) M. Morton and L. J. Fetters, *J. Polymer Sci.,* **A2,** 331 (1964).

(71) S. Bywater and D. J. Worsfold, *Can. J. Chem.,* **81,** 2748 (1959).

(72) D. J. Cram and K. R. Kopecky, *J. Am. Chem. Soc.,* **81,** 2748 (1959).

(73) D. J. Cram, *J. Chem. Educ.,* **37,** 317 (1960).

(74) C. E. H. Bawn and A. Ledwith, *Quart. Rev. (London),* **16,** 361 (1962).

CHAPTER III

The Observation of Polymer Chain Conformation by NMR

1. *Model Compound Conformations.* In Chapter I, we described some of the symmetry properties of small model compounds—the 2,4-disubstituted pentanes and 2,4,6-trisubstituted heptanes—which serve as models for vinyl polymer chains. In this chapter, we discuss the interpretation of the NMR spectra of 2,4-disubstituted pentanes in terms of *chain conformation*, and then apply the results to the particular case of the 2,4-diphenylpentanes, which are the analogs of polystyrene. The discussion follows the general lines laid down by Doskocilova *et al.*,[1,2] McMahon and Tincher,[3] Bovey *et al.*,[4] and Shimanouchi *et al.*[5]

2. *The Averaging of Vicinal Couplings.* It is well recognized that under conditions of rapid equilibration, *i.e.,* when the characteristic time of bond rotation is substantially smaller than the reciprocal of the spectral line spacings (expressed in cps), the observed vicinal couplings in substituted ethanes are an average over the conformations present:

$$J_{\text{obs}} = \Sigma_i X_i J_i \qquad \text{(III-1)}$$

where X_i and J_i represent the mole fraction and vicinal coupling of each conformer. This expression holds for ethanes of the type $X_2CH \cdot CH_2Y$ or $X_2CH \cdot CHY_2$, but for ethanes of the type XCH_2CH_2Y there are two observable couplings unless the populations of the two mirror-image *gauche* conformers and of the *trans* conformer are all equal. For ethanes with an asymmetric carbon atom, $XCH_AH_BCH_CYZ$, a parallel expression applies to each of the distinguishable protons H_A and H_B. (See Chapter I, Sec. 1.)

For *meso*-2,4-disubstituted pentanes, there are nine possible conformers, including three pairs of equi-energy mirror-image conformers (4, 5, and 6 in Fig. III-1) which are con-

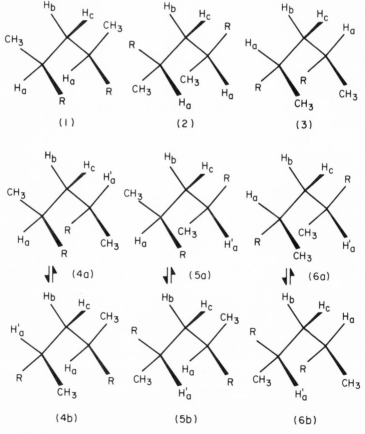

Fig. III-1. Staggered conformers of *meso*-2,4-disubstituted pentanes.

vertible into each other, *i.e.,* 4*a* into 4*b*, 5*a* into 5*b*, etc., by 120° rotations about the bonds between the α- and β-carbons. Conformers 1, 2, and 3 have a plane of symmetry; the "mirror-image" conformers, 4, 5, and 6, have no plane or axis of symmetry.

For *racemic*-2,4-disubstituted pentanes (Fig. III-2), conformers 1, 2, and 3 have twofold symmetry axes; the other three, 4, 5, and 6, are analogous to conformers 4, 5, and 6 of the *meso* isomer in having no plane or axis of symmetry, but

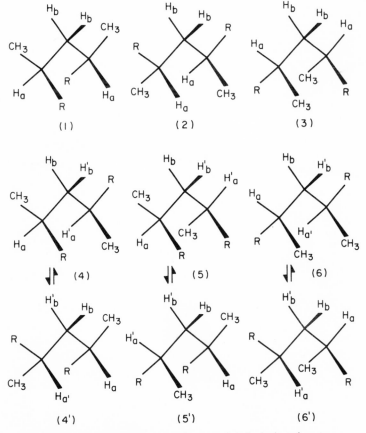

Fig. III-2. Staggered conformers of *racemic* 2,4-disubstituted pentanes.

differ in being converted into identical forms ("identomers"), 4', 5', and 6', rather than mirror images, by 120° rotations about the α, β bonds. This isomer can of course be converted to its mirror image only by inversion at both asymmetric carbon atoms. Only one of the two optical isomers is represented in Fig. III-2, since enantiomers are normally indistinguishable by NMR.

In the *meso*-2,4-disubstituted pentanes, the β-methylene protons H_B and H_C are in different environments in all conformations, and do not exchange environments upon conformational equilibration between mirror-image forms, as one can readily see from an inspection of models or from Fig. III-1. (It should be noted, however, that H_A and $H_{A'}$ are always equivalent, either by symmetry or by conformational averaging.) H_B and H_C will thus (barring fortuitous equivalence) always have different chemical shifts, as we have seen in Chapter I, and will in principle experience different couplings to the α-protons. In the asymmetrical conformers 4, 5, and 6, the couplings experienced by H_B and H_C to the α-protons will each be averaged by conformational equilibration between pairs of equi-energy mirror images. These considerations lead to the following expressions:

$$J_{AB} = X_1 J_{1\text{-AB-}g} + X_2 J_{2\text{-AB-}t} + X_3 J_{3\text{-AB-}g}$$
$$+ X_4 \bar{J}_{4\text{-AB}} + X_5 \bar{J}_{5\text{-AB}} + X_6 \bar{J}_{6\text{-AB}} \qquad \text{(III-2}a\text{)}$$

$$J_{AC} = X_1 J_{1\text{-AC-}t} + X_2 J_{2\text{-AC-}g} + x_3 J_{3\text{-AC-}g}$$
$$+ X_4 \bar{J}_{4\text{-AC}} + X_5 \bar{J}_{5\text{-AC}} + X_6 \bar{J}_{6\text{-AC}} \qquad \text{(III-2}b\text{)}$$

where the X_i of course include both mirror-image forms in conformers 4, 5, and 6, and the subscripts g and t denote *gauche* and *trans* couplings. (The designations of H_B and H_C here and in Fig. III-1 are of course arbitrary and could be interchanged.) In the asymmetric conformers, the averaged couplings $\bar{J}_i$ are given by

$$\bar{J}_{4\text{-AB}} = \tfrac{1}{2} (J_{4\text{-AB-}g} + J'_{4\text{-AB-}g}) \qquad \text{(III-3}a\text{)}$$

$$\bar{J}_{4\text{-AC}} = \tfrac{1}{2} \left(J_{4\text{-AC-}t} + J_{4\text{-AC-}g} \right) \tag{III-3b}$$

$$\bar{J}_{5\text{-AB}} = \tfrac{1}{2} \left(J_{5\text{-AB-}g} + J_{5\text{-AB-}t} \right) \tag{III-4a}$$

$$\bar{J}_{5\text{-AC}} = \tfrac{1}{2} \left(J_{5\text{-AC-}t} + J_{5\text{-AC-}g} \right) \tag{III-4b}$$

$$\bar{J}_{6\text{-AB}} = \tfrac{1}{2} \left(J_{6\text{-AB-}g} + J_{6\text{-AB-}t} \right) \tag{III-5a}$$

$$\bar{J}_{6\text{-AC}} = \tfrac{1}{2} \left(J_{6\text{-AC-}g} + J'_{6\text{-AC-}g} \right) \tag{III-5b}$$

Primes are employed to distinguish *gauche* couplings within the same conformer which are in principle nonequivalent, although probably in fact nearly equal. Assuming equality of all *gauche* and of all *trans* couplings, we can then rewrite eqs. (III-2a) and (III-2b) as

$$J_{AB} = X_1 J_g + X_2 J_t + X_3 J_g + X_4 J_g$$
$$+ X_5 (J_g + J_t)/2 + X_6 (J_g + J_t)/2 \tag{III-2c}$$

$$J_{AC} = X_1 J_t + X_2 J_g + X_3 J_g + X_4 (J_g + J_t)/2$$
$$+ X_5 (J_g + J_t)/2 + X_6 J_g \tag{III-2d}$$

In the *racemic* conformers 1, 2, and 3, the β-methylene protons H_B are identical by symmetry, but experience two different observable couplings to the α protons. In conformers 1 or 2 these would have the magnitudes expected of *trans* and *gauche* vicinal couplings if one conformer or the other were present alone. In the unsymmetrical conformers 4, 5, and 6, rapid equilibration between "identomers" averages the environments of H_B and $H_{B'}$ (and of H_A and $H_{A'}$), but two different couplings to the α-protons would still be observable (at least in principle) for each conformer, supposing again that it alone were present. The observed couplings will be given by

$$J_{AB} = X_1 J_{1\text{-}t} + X_2 J_{2\text{-}g} + X_3 J_{3\text{-}g} + X_4 (J_{4\text{-}g} + J_{4\text{-}t})/2 +$$
$$X_5 (J_{5\text{-}g} + J_{5\text{-}t})/2 + X_6 (J_{6\text{-}g} + J'_{6\text{-}g})/2 \tag{III-6a}$$

$$J'_{AB} = X_1 J_{1\text{-}g} + X_2 J_{2\text{-}t} + X_3 J'_{3\text{-}g} + X_4 (J'_{4\text{-}g} + J_{4\text{-}t})/2 +$$
$$X_5 (J_{5\text{-}g} + J'_{5\text{-}g})/2 + X_6 (J_{6\text{-}g} + J_{6\text{-}t})/2 \tag{III-6b}$$

Assuming equality of all *gauche* and of all *trans* couplings, as before, we can simplify these equations:

$$J_{AB} = X_1 J_t + X_2 J_g + X_3 J_g + X_4 (J_g + J_t)/2 +$$
$$X_5 (J_g + J_t)/2 + X_6 J_g \qquad \text{(III-6}c\text{)}$$

$$J'_{AB} = X_1 J_g + X_2 J_t + X_3 J_g + X_4 (J_g + J_t)/2 +$$
$$X_5 J_g + X_6 (J_g + J_t)/2 \qquad \text{(III-6}d\text{)}$$

which closely parallel eqs. (III-2c) and (III-2d) for the *meso* isomer.

If there were no enthalpy differences among the conformers (or if the temperature were very high and there were no entropy differences other than those corresponding to the statistical weights of conformers 4, 5, and 6), it follows from the above that the two couplings J_{AB} and J'_{AB} would be equal. This is also possible under less stringent conditions, as will be seen below. It cannot yet be stated whether J_{AB} or J'_{AB} will be the larger, since the assignment is arbitrary. Some independent indication of the conformer populations is required.

3. *Conformer Populations*

a. Meso *2,4-Disubstituted Pentanes.* It is evident from eqs. (III-2a) and (III-2b) or from (III-2c) and (III-2d) that in the *meso* spectrum J_{AB} and J_{AC} will differ substantially if either conformer 1 or conformer 2 is strongly preferred. If conformer 3 is preferred, the couplings will be nearly equal and will be of the order of 2-3 cps. In Fig. III-3 are shown the spectra of *meso*-2,4-diphenylpentane at 35° in carbon tetrachloride and at 180° in *o*-dichlorobenzene. The α-proton spectrum appears as a sextet at lower field, and the β-CH$_2$ spectrum as ten or eleven distinguishable peaks at higher field. Analysis of these spectra as AA'BCDD' spin systems (see reference 4 for a detailed discussion) was carried out by comparison with computer-simulated spectra, shown in Fig. III-3. It was found that J_{AB} and J_{AC} are equal within experimental uncertainty and approximately 7.4 cps near room temperature, decreasing slightly but remaining equal at elevated temperatures. (It should be noted that for other *meso*-2,4-

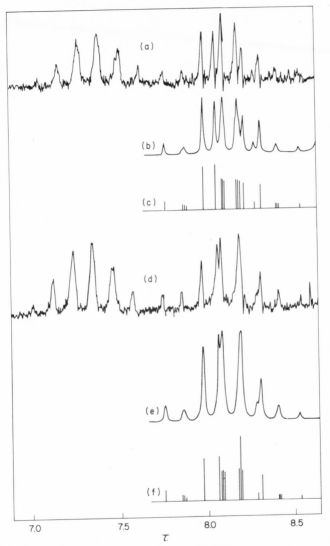

Fig. III-3. Spectrum of *meso*-2,4-diphenylpentane: (*a*) in carbon tetrachloride at 35°; (*b*) calculated spectrum with linewidth = 0.7 cps; (*c*) "stick" spectrum corresponding to (*b*); (*d*) in *o*-dichlorobenzene at 180°; (*e*) calculated spectrum with linewidth 0.8 cps; (*f*) "stick" spectrum corresponding to (*e*).

disubstituted pentanes, J_{AB} is observed to exceed J_{AC} by as much as 1 cps; we shall discuss this point later.) The symmetric conformers therefore cannot be present in significant proportion, a conclusion which is compatible with simple steric considerations, since the nonbonded repulsive interactions between phenyl and methyl groups will certainly tend to exclude these forms. Of the asymmetric conformers, 4, 5, and 6, steric considerations, supported by nonbonded interaction-energy calculations,[3,6,7] indicate that in general conformer 5 will be very strongly preferred. This is in agreement with the observed couplings. We then have from eqs. (III-2a) and (III-2b)

$$J_{AB} - J_{AC} = \bar{J}_{5\text{-}AB} - \bar{J}_{5\text{-}AC}$$

which from eqs. (III-4a) and (III-4b) becomes

$$J_{AB} - J_{AC} = (\Delta J_g + \Delta J_t)/2$$

where

$$\Delta J_g = J_{5\text{-}AB\text{-}g} - J_{5\text{-}AC\text{-}g}$$

and

$$\Delta J_t = J_{5\text{-}AB\text{-}t} - J_{5\text{-}AC\text{-}t}$$

Thus, if only conformer 5 is present, equal vicinal couplings will be observed if ΔJ_g and ΔJ_t are zero (or if their sum is zero). It seems very probable that both ΔJ_g and ΔJ_t will in fact be nearly zero in compounds of low polarity such as the diphenylpentanes, since there is little reason to expect the dihedral angles to differ greatly for protons H_B and H_C. For the dichloride, dibromide, and diol, McMahon and Tincher[3] have reported values of $|J_{AB} - J_{AC}|$ of the order of 0.5 to 1.0 cps and attribute the difference to deviations from exact staggering. Doskocilova and Schneider[2] do not distinguish the *meso* vicinal couplings in the dichloride and diacetate, although in a previous paper Doskocilova[1] reported two appreciably different couplings for the dichloride. More recently, Schneider et al.[8] have suggested that appreciable deviations from exact staggering, which could lead to marked deviations from expected coupling values, may occur. This will be discussed later.

b. Racemic *2,4-diphenylpentane.* The spectra of *racemic* 2,4-diphenylpentane at a number of temperatures are shown in Fig. III-4. At lower temperatures, the β-CH$_2$ resonance appears not as the triplet which might be expected but as a doublet of doublets. This part of the spectrum is very temperature-dependent (as reported by McMahon and Tincher[3] for the dihalopentanes), and becomes triplet-like at high temperature. Analysis of the spectra shows that this isomer exhibits two widely different vicinal couplings, J_{AB} and J'_{AB} at lower temperatures, but that as the temperature is increased the difference between J_{AB} and J'_{AB}, designated by δJ, decreases. For example, at $-50°$ (CS$_2$), $\delta J = 8.0$ cps; at $35°$ (carbon tetrachloride), $\delta J = 5.0$ cps; at $100°$ (o-dichlorobenzene) it decreases to 3.3 cps and finally to 1.0 cps at $200°$. These observations find a ready explanation in terms of eqs. (III-6c) and (III-6d). The large value of δJ at lower temperatures indicates that neither conformer 3 nor 4 can predominate, since this would lead to a small or zero value. A preference for conformer 5 or 6 might explain the data, but it would be difficult to account for a value of δJ as large as 8 cps if J_t and J_g have the usual values; more significantly, these conformers appear to be strongly excluded from steric considerations.[3-5] We may then rewrite eqs. (III-6c) and (III-6d), eliminating all terms in conformers other than 1 and 2

$$J_{AB} = X_1 J_t + X_2 J_g \qquad \text{(III-7a)}$$

$$J'_{AB} = X_1 J_g + X_2 J_t \qquad \text{(III-7b)}$$

from which it is evident that δJ will approach zero as X_1 and X_2 approach equality, since

$$| \delta J | = (X_1 - X_2)(J_t - J_g) \qquad \text{(III-8)}$$

We also have

$$J_{AB} + J'_{AB} = J_t + J_g \qquad \text{(III-9)}$$

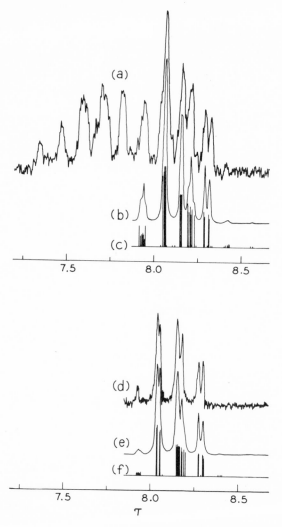

Fig. III-4. Spectrum of *racemic* 2,4-diphenylpentane: (*a*) in carbon disulfide at −50°; (*b*) calculated spectrum with linewidth = 0.6 cps; (*c*) "Stick" spectrum corresponding to (*b*); (*d*) in carbon tetrachloride at 35°; (*e*) calculated spectrum with linewidth = 0.7 cps; (*f*) "stick" spectrum corre-

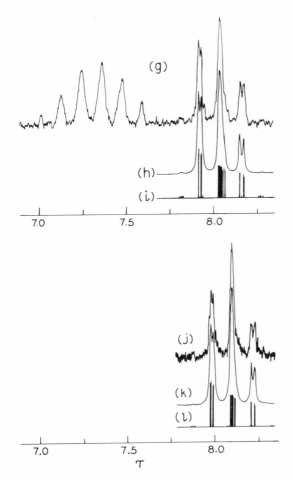

sponding to (*e*); (*g*) In *o*-dichlorobenzene at 100°; (*h*) calculated spectrum with linewidth = 0.8 cps; (*i*) "stick" spectrum corresponding to (*h*); (*j*) in *o*-dichlorobenzene at 200°; (*k*) calculated spectrum with linewidth = 0.8 cps; (*l*) "stick" spectrum corresponding to (*k*).

and it is in fact found experimentally that the quantity $(J_{AB} + J'_{AB})$ is nearly independent of temperature, as would be expected. If it is assumed that all *gauche* couplings are equal and that all *trans* couplings are equal in both the *meso* and *racemic* isomers, and that the *meso* isomer is entirely conformer 5, it further follows that

$$J_g + J_t = (J_{AB} + J'_{AB})_{racemic} = (J_{AB} + J_{AC})_{meso} \quad \text{(III-10)}$$

This relationship is also obeyed within experimental error at all temperatures. These observations are unlikely to be merely fortuitous and provide support for the assumptions which must necessarily be made in order to draw any useful conclusions. $J_g + J_t$ is thus found to equal 14.8 ± 0.2 cps. If we define $J_t - J_g$ as ΔJ and assume that δJ is positive— *i.e.*, that conformer 1 predominates over the temperature range studied—it is readily shown that

$$(G_2 - G_1)_{rac} = \Delta G_{rac} = -RT \ln\left(\frac{\Delta J - \delta J}{\Delta J + \delta J}\right) \quad \text{(III-11)}$$

$$(H_2 - H_1)_{rac} = \Delta H_{rac} = -R\left[\partial \ln\left(\frac{\Delta J - \delta J}{\Delta J + \delta J}\right)/\partial(1/T)\right]_P \quad \text{(III-12)}$$

and

$$(S_2 - S_1)_{rac} = \Delta S_{rac} = -(\partial \Delta G_{rac}/\partial T)_P \quad \text{(III-13)}$$

From the slope and intercept of an Arrhenius plot of the coupling data according to eq. (III-12), it is found that

$$\Delta H_{rac} = 1.66 \pm 0.20 \,\text{kcal}$$

$$\Delta S_{rac} = 3.2 \pm 0.5 \,\text{e.u.}$$

$$\Delta G_{rac}(298°) = 0.70 \pm 0.20 \,\text{kcal}$$

For this purpose, a value of ΔJ was chosen (10.0 cps) such that a linear plot was obtained.

It is seen that conformer 2 is favored by entropy, possibly because the phenyl groups are farther apart and can rotate more freely. (Such an effect would be absent for atomic substituents.) At temperatures above *ca.* 520°K conformer 2 will predominate.

In general, similar conclusions concerning conformational preferences have been reached for other 2,4-disubstituted pentanes; detailed references to this work are given in Chapter I. Exceptional behavior is shown by the 2,4-dihydroxypentanes, in which hydrogen bonding between the hydroxyl groups stabilizes the conformations which would otherwise tend to be excluded.

Schneider *et al.*,[8] as indicated above, have recently refined their analysis of the NMR and infrared spectra of the 2,4-dichloropentanes to take account of the fact that in these more polar compounds J_{AB} and J_{AC} are not quite equal for the *meso* compound and that eqs. (III-9) and (III-10) do not hold, temperature dependence being appreciable. They also observed small infrared peaks apparently corresponding to conformers previously excluded on energy grounds. They showed that if both torsional and nonbonded potentials are included, such forms as 4 for the *meso* isomer (see Fig. III-1) and 5 for the *racemic* isomer (Fig. III-2) may be allowed in small proportion, but with dihedral angles differing by several degrees from exact staggering.

4. *2,4,6-Trisubstituted Heptanes.*   The extension of model compound studies to the 2,4,6-trisubstituted heptanes has been discussed in general terms in Chapter I, and references to this work are given there. The most carefully studied compounds are the three trichloro stereoisomers (see p. 7), the analogs of polyvinyl chloride.[5,9,10] The analysis of the NMR spectra follows the general lines described for the pentanes in the previous section, but with certain complications. For example, the protons H_A and $H_{A'}$ in the syndiotactic isomer (p. 8) are nonequivalent, as we have already indicated. It is also found that the difference in chemical shift between the protons H_A and H_B in the isotactic isomer is markedly smaller (*ca.* 0.10 p.p.m.) than in the *meso*-dichloropentane (*ca.* 0.25 p.p.m.), illustrating a significant shielding influence of distant groups which could be misleading if the pentane model were interpreted too literally as a guide to the

polymer spectrum (see p. 21 *et seq*.). Again, the heterotactic isomer offers some complications not present in the pentane models; for example, the "*meso*" portion of this isomer cannot be assumed to take on both "mirror-image" conformations with equal probability as does the *meso*-pentane (Fig. III-1) because the molecular environment is unsymmetrical.

The conclusions as to conformational preferences of the trichloroheptanes, as deduced by Doskocilova *et al.*[10] from NMR and infrared studies, are summarized in Table III-1

TABLE III-1. Conformational Preferences (in mole fractions) of 2,4,6-Trisubstituted Heptanes at *ca*. 300°K (refs. 5, 10, 11, 13)

Isomer	Conformer	Substituent		
		Cl	C_6H_5	CO_2CH_3
isotactic	TGTG	0.80	0.63	0.80
	GTTG	0.20	0.37	0.20
syndiotactic	TTTT	0.85	0.66	0.38
	GGTT	0.15	0.34	0.62
heterotactic	TTTG	*ca*. 0.55	0.44	0.30
	TTGT	*ca*. 0.45	0.40	0.43
	GGTG	< 0.05	0.16	0.27

and Fig. III-5. (The conclusions of Abe *et al.*[9] are generally in good agreement, but they do not attempt to give quantitative estimates of minor components.) The syndiotactic isomer strongly prefers the planar zigzag conformation. The isotactic isomer strongly prefers the TGTG or GTGT conformation, as might be anticipated from the behavior of the *meso*-pentane. As we shall see, both conformations are segments of 3_1 helices (*i.e.*, helices that have 3-fold symmetry and repeat in one turn). Agreement is less satisfactory concerning the heterotactic isomer, which Abe *et al.*[9] believe to be predominantly GTTT.

Pivcova *et al.*[11] have analyzed the NMR spectra of the 2,4,6-triphenylheptanes described by Lim *et al.*[12] The conformational preferences at 300°K are shown in Table III-1.

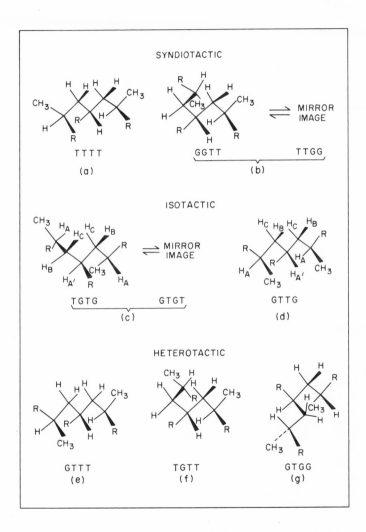

Fig. III-5. Allowed conformations of 2,4,6-trisubstituted heptanes (refs. 8 and 9).

They are similar to those of the chloro-compounds, but less marked, the minor conformations being more prominent. Doskocilova *et al.*[13] have also reported on the 2,4,6-tricarbo-methoxyheptanes, the polymethyl acrylate analogs, and these results are included in Table III-1. The preferences of the isotactic and heterotactic isomers are similar to those of the corresponding triphenyl compounds, but the syndiotactic isomer actually prefers the folded GGTT structure. For the syndiotactic 2,4,6-triphenylheptane,[11] this structure is favored by entropy (by more than the 1.4 e.u. arising from the existence of equi-energy mirror-image forms), and so its contribution tends to increase with temperature, as does the GG form of the *racemic*-2,4-diphenylpentane. It is probably reasonable to expect similar behavior from the trichloro- and tricarbo-methoxy- compounds, and so one must assume that such forms may contribute to polymer chain conformations.

Flory[14] has presented a general theory of stereochemical equilibrium which, among other things, predicts the conformer populations of the tricarbomethoxyheptanes within the probable experimental error, using two adjustable parameters.

5. *Polymer Chain Conformation.* Conventional measurements of the conformations of polymer molecules in solution, such as light scattering and viscosity, give information concerning the overall extension of the chains in solution, conveniently expressed by the mean square end-to-end distance $<r^2>_0$. In contrast, spectroscopic observations, NMR in particular, give information concerning local conformational preferences. This approach is, in principle at least, a more effective one, for if the local behavior is known with sufficient precision, then the overall chain extension is likewise known, provided we have some measure of the molecular weight. There are difficulties, however. We have seen that the analysis of the NMR spectra of model compounds in terms of conformer populations involves the assumption of exact staggering or (which is nearly the same thing) the assumption that J_g and J_t are the same in all conformers. It appears that this assumption may be too stringent,[8,15] and that

dihedral angles in *gauche* conformations even of nonpolar chains such as those of alkanes[15] or diphenylpentanes may depart from the exact staggering values. The apparent self-consistency of the NMR analysis of the diphenylpentanes may result in part from the fact that the function describing the dependence of vicinal coupling on dihedral angle (assumed to be that given by Karplus[16]) is insensitive near 180° and not maximally sensitive near 60°. The only way to evade this difficulty would seem to be the direct observation of the conformers and their couplings by "freezing" them at low temperatures, but at present this is experimentally feasible for open-chain compounds only if highly substituted by halogens and other large groups, which raise the rotational barriers to *ca.* 8–10 kcal.

Even accepting the simplifying assumption of equal values of J_g and of J_t in all conformations, we still encounter difficulties in the interpretation of polymer NMR spectra, mainly because of their relatively poor resolution. Thus, for example, the spectra of syndiotactic polymers exhibit a triplet β-methylene proton resonance, never the four- or five-line spectra (plus weak outer peaks) of Fig. III-4. This does not necessarily mean that TT and GG conformations are about equally populated, for the spectra are of the "deceptively simple" sort, as described by Abraham and Bernstein.[17] The vicinal couplings J_{AB} and J'_{AB} may differ markedly and yet the triplet appearance is maintained. The criterion for such a spectrum to be deceptively simple is that[17]

$$\delta\nu \geq \delta J^2 / 2 J_{\text{gem}} \qquad \text{(III-14)}$$

where $\delta\nu$ is the spectral linewidth and δJ is $|J_{AB} - J'_{AB}|$, as previously defined. We see that even if δJ is as large as *ca.* 7 cps, the spectrum will still be deceptive, and therefore not as informative as one might hope. Some information can be obtained, however, by careful computer simulation of the observed triplet, for as can be seen in Fig. III-4, the height of the central peak is dependent upon δJ, as are the intensities of the small satellite peaks on each side. So far, no syndiotactic polymer has been critically examined in this manner. Studies of model compounds, particularly the 2,4,6-trisubstituted heptanes, indicate that TT conformations may predominate but that, as we have seen in Sec. 4, the conformer preference is strongly dependent on the nature of the side-chain.

Up to the present, only a very limited effort has been made to draw conclusions concerning polymer chain conformations from NMR spectra. The following cases have been considered:

a. Polystyrene. It is found experimentally for the 2,4-diphenylpentanes that (see also Sec. 3 of this chapter and Fig. I-14)

$$J_{AB} \quad = \quad J_{AC} \quad = \quad J_{AB} \quad = \quad J_{AC} \quad = 7.10 \pm 0.10$$

| (*meso* pentane) | (*meso* pentane) | (isotactic polystyrene) | (isotactic polystyrene) |

Again, there is considerably greater margin for error in the determination of the couplings of the polymer than of the pentanes because of the greater linewidth of the polymer spectrum. However, the values given are probably correct with ± 0.2 cps, as a wider variation produces a simulated spectrum which departs observably from the experimental. It seems reasonable to conclude that the TG conformation (conformer 5 in Fig. III-1) is strongly preferred by the polymer as well as the model compound. As we have seen, this is a segment of the 3_1 helical conformation that the polymer is believed to have in the crystalline state[18-20] (Fig. III-6a). The corresponding heptane conformation, TGTG (or GTGT), is preferred (Fig. III-5), but the GTTG conformation, representing the junction of two helices of opposite sign, is present to an extent that apparently cannot be ignored (Table III-1). (It should again be pointed out that the parallel studies of Abe et al.[9] and Doskocilova et al.[10] on isotactic 2,4,6-trichloroheptane led to different conclusions as to the presence of the GTTG conformation; the spectral interpretation of Abe et al. was inconsistent with any significant population of this form. Thus, a question remains, particularly in view of the relatively

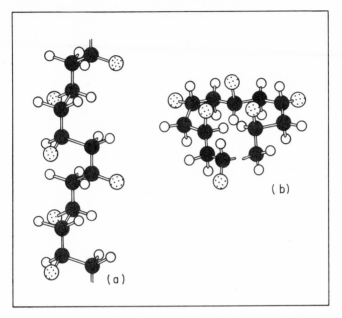

Fig. III-6. (a) Isotactic vinyl polymer chain in 3_1 helical conformation; darkest balls denote carbon backbone, small unshaded balls are hydrogen atoms, large shaded balls are α-1 substituents; (b) Same chain as (a) and with same local conformation with respect to CH_2 groups, but with alternating sign of mirror-image conformations along the chain.

poor resolution of the triphenylheptane spectra.) The GTTG sequences, if present in the polymer chain, are necessarily short because of steric requirements; as can be seen in Fig. III-6b, a chain composed of such sequences, ···GTTGGTTGGTTG···, quickly returns to its starting point. The presence of such sequences in substantial proportion also complicates the coupling patterns, and makes the six-spin "dimer" model (p. 27) no longer a valid approximation of the polymer chain.

The NMR evidence thus appears to be consistent with a strong predominance of 3_1 helical sequences in isotactic polystyrene chains in solution. There is also optical evidence for such helical conformations in both solution and melt.[21-24] It is evident, however, that some other conformations must be present. Viscosity[25] and light-scattering[26] measurements indicate that the mean square end-to-end distance $\langle r^2 \rangle_0$ in good solvents is little if any larger for isotactic polystyrene than for "atactic" (actually predominantly syndiotactic; see p. 29) polystyrene, whereas it would obviously be very much greater if only $\cdots TGTG \cdots$ were present. If this picture is correct, the isotactic chains may deviate widely from the 3_1 helical conformation by undergoing mirror inversions ($\cdots TG \cdots \rightleftharpoons \cdots GT \cdots$); the microbrownian motion of the chain in solution may be thought of as inversions of this sort moving along the chain in a random fashion. It is implied by some authors[27] that, in an isotactic chain, only one junction between left- and right-handed 3_1 helices, *i.e.*,

$$(GT)_n(TG')_m$$

(the TG' denoting a conformation of opposite sense to GT), is energetically permissible. It appears, nevertheless, that, at least so far as immediate local steric requirements are concerned, conformations similar to (*b*), but not returning completely upon themselves, should be allowed, and that therefore many helical segments of opposite sign are permissible in a single chain. There is also the possibility that syndiotactic sequences may be present in sufficient proportion to reduce $\langle r^2 \rangle_0$ to its unexpectedly low value. We shall consider this point further below.

b. Polyacrylates. Isotactic polyacrylate chains are known to have a 3_1 helical conformation in the crystalline state.[18,28] Yoshino *et al.*[29] have prepared isotactic polymers, using anionic initiators, from the monomers

$$\begin{array}{c} \text{H} \qquad\qquad \text{H} \\ \diagdown \qquad\quad \diagup \\ \text{C}=\text{C} \qquad\qquad (a) \\ \diagup \qquad\quad \diagdown \\ \text{ROOC} \qquad\qquad \text{D} \end{array}$$

and

$$\begin{array}{c} \text{H} \qquad\qquad \text{D} \\ \diagdown \qquad\quad \diagup \\ \text{C}=\text{C} \qquad\qquad (b) \\ \diagup \qquad\quad \diagdown \\ \text{ROOC} \qquad\qquad \text{H} \end{array}$$

The mole ratio of the *trans* (*b*) to *cis* (*a*) monomer was 3:2; R was —CH_3 and —$CH(CH_3)_2$. The disposition of *cis* and *trans* monomer units along the polymer chains was believed to be random, but whether this is the case or not or whether addition to the double bond is "*cis*" or "*trans*" (see Chapter II) does not affect the conclusions. Replacement of one of the β-methylene protons by deuterium simplified the spectrum, and allowed a less ambiguous estimate of the vicinal couplings to be made. The spectrum of the polymethyl β-d_1-acrylate in methyl formate solution is shown in Fig. III-7.

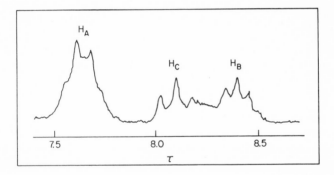

Fig. III-7.   Spectrum of the α and β protons of isotactic polymethyl β-d_1-acrylate, observed in methyl formate [from T. Yoshino *et al.*, *J. Phys. Chem.*, 70, 1059 (1966)].

In addition to the α-proton resonance centered at *ca.* 7.6τ (and clearly not a simple triplet), triplet resonances of the *erythro* and *threo* β-protons (see p. 19) appear at *ca.* 8.1 τ and 8.4 τ, respectively. The significant feature of the spectrum is that these triplets do not have the same spacing, the *erythro* protons being more strongly coupled to the α-protons (8.1 cps) than the *threo* protons (5.6 cps).* Similar spectra were obtained in chloroform and benzene. If exact staggering and

---

*NOTE ADDED IN PROOF: More recent studies (see page 32) have confirmed that these couplings do indeed differ in magnitude, though not so greatly as reported by Yoshino *et al.*

symmetrical Karplus functions are assumed and if the chain is in the 3_1 helical conformation, these couplings should be the same. Actually, the validity of one or both of these assumptions is already called into question by the observation that in *meso*-dichloropentane and *meso*-dicarbomethoxypentane, which almost certainly are TG, the *threo* and *erythro* couplings (protons B and C, respectively, in Figs. III-1 and III-2) differ, the latter being about 1.0 cps larger. Substantial populations of conformations normally considered to be very strongly excluded have been suggested by Yoshino[29] to explain the discrepancy. Conformations such as indicated in Fig. III-6*d* would not have the effect of making the *erythro* coupling appear larger than the *threo* coupling. The most probable explanation at present appears to be that the dihedral angles in the polyacrylate chain depart from exact staggering values somewhat more than do those of the model compounds, and that consequently the predominant chain conformation is no longer exactly a 3_1 helix.

Flory, Mark, and Abe[30] have calculated the characteristic ratios, $\langle r^2 \rangle_0 / nl^2$, *i.e.,* the ratio of the actual mean square end-to-end distance to that expected for a random flight chain of n links of length l, for vinyl polymer chains in which P_m (see Chapter I) varies from 0 (syndiotactic) to 1 (isotactic). The ratio is rather insensitive at lower values of P_m, remaining near 10 until P_m exceeds about 0.9, after which it shoots rapidly up toward a very high value characteristic of the $\cdots$ GTGT $\cdots$ helix. The characteristic ratio is found experimentally to be insensitive to P_m even when P_m appears to approach 1. Mark et al.[31,32] find characteristic ratios of 9.7, 7.1, and 7.2 for "isotactic," atactic, and syndiotactic polyisopropyl acrylates, respectively. Flory et al.[30,33] resolve this disagreement by proposing that P_m is in fact not so high as is estimated. This point has not yet been resolved experimentally. Clearly, a more searching measurement of the stereoregularity of the isotactic polymer is needed. Measurements at 220 MHz may supply this.

The calculations of Flory et al.[30] do not allow substantial contributions of conformations such as shown in Fig. III-6*b*.

If in fact a polymer having a P_m very nearly 1.0 proves to have a characteristic ratio of nearly 10, revision of conformational energy calculations will be called for.

c. Polyethylene Oxide. The conformation of polyethylene oxide in aqueous and chloroform solutions and in the melt has been investigated by Connor and McLauchlan.[34] Because all the protons have the same chemical shift under conditions of rapid rotational isomerization, it was necessary to observe the C^{13} sidebands, which appear as a doublet straddling the main proton resonance. Each member of the doublet is complex and may be treated as half of the AA′BB′ spectrum of the monomer unit

$$[C^{13}H_2CH_2\!-\!O]_n$$

The C^{13} satellite spectrum closely resembles that of dioxane.[35] Analysis of the spectrum yields values of the two vicinal coupling constants characteristic of such systems. These, however, are averages and must be treated in a manner parallel to that already discussed in connection with model pentane and heptane spectra. It can be shown that, making the usual assumptions of constancy of J_g and J_t

$$J + J' = \tfrac{1}{2}[J_t + 3J_g + X_t(J_t - J_g)] \qquad \text{(III-15)}$$

where J and J' are the two vicinal coupling constants, the sum of which is obtained directly from the spectrum; X_t is the mole fraction of *trans* conformer:

It is found that $J + J'$ increases with temperature. Equation (III-15) indicates that the *gauche* conformer must then be of lower energy. The dependence of $J + J'$ upon temperature is actually greater than can be accounted for by a single reasonable value for the energy difference. This approach is a potentially interesting one but must be developed further before the results can be reliably interpreted.

REFERENCES FOR CHAPTER III

(1) D. Doskocilova, *J. Polymer Sci.*, **B2**, 421 (1964).

(2) D. Doskocilova and B. Schneider, *Coll. Czech. Chem. Communs.*, **29**, 2290 (1964).

(3) P. E. McMahon and W. C. Tincher, *J. Mol. Spectry.*, **15**, 180 (1965).

(4) F. A. Bovey, F. P. Hood, E. W. Anderson, and L. C. Snyder, *J. Chem. Phys.*, **42**, 3900 (1965).

(5) T. Shimanouchi, M. Tasumi, and Y. Abe, *Makromol. Chem.*, **86**, 43 (1965).

(6) T. Shimanouchi and M. Tasumi, *Spectrochim. Acta*, **17**, 775 (1961).

(7) P. E. McMahon, *Trans. Faraday Soc.*, **61**, 197 (1965).

(8) B. Schneider, J. Stokr, D. Doskocilova, S. Sykora, J. Jakes, and M. Kolinsky, Preprints IUPAC Macromolecular Symposium, Brussels, 1967; Sec. IX.

(9) Y. Abe, M. Tasumi, T. Shimanouchi, S. Satoh, and R. Chujo, *J. Polymer Sci.*, **4A1**, 1, 13 (1966).

(10) D. Doskocilova, J. Stokr, B. Schneider, H. Pivcova, M. Kolinsky, J. Petranek, and D. Lim, *J. Polymer Sci.*, **C16**, 215 (1967).

(11) H. Pivcova, M. Kolinsky, D. Lim, and B. Schneider, Preprints IUPAC Macromolecular Symposium, Brussels, 1967.

(12) D. Lim, M. Kolinsky, J. Petranek, and D. Doskocilova, *J. Polymer Sci.*, **4B**, 645 (1966).

(13) D. Doskocilova, S. Sykora, H. Pivcova, B. Obereigner, and D. Lim, Preprints IUPAC Macromolecular Symposium, Tokyo-Kyoto, 1966.

(14) P. J. Flory, *J. Am. Chem. Soc.*, **89**, 1798 (1967).

(15) A. Abe, R. L. Jernigan, and P. J. Flory, *J. Am. Chem. Soc.*, **88**, 631 (1966).

(16) M. Karplus, *J. Chem. Phys.*, **30**, 11 (1959).

(17) R. J. Abraham and H. J. Bernstein, *Can. J. Chem.*, **39**, 216 (1961).

(18) G. Natta, *Makromol. Chem.*, **35**, 94 (1960).

(19) G. Natta, P. Corradini, and I. W. Bassi, *Nuovo Cimento*, (suppl. 1), **15**, 68 (1960).

(20) R. L. Miller and L. E. Nielsen, *J. Polymer Sci.*, **55**, 743 (1961).

(21) M. Takeda, K. Iimura, A. Yamada, and Y. Imamura, *Bull. Chem. Soc. Japan*, **32**, 1151 (1959).

(22) T. Onishi and S. Krimm, *J. Appl. Phys.*, **32**, 2320 (1961).

(23) M. T. Vala, Jr., and S. A. Rice, *J. Chem. Phys.*, **39**, 2348 (1963).

(24) J. W. Longworth, *Biopolymers*, **4**, 1131 (1966).

(25) F. Danusso and G. Moraglio, *Makromol. Chem.,* **38,** 250 (1958).

(26) W. R. Krigbaum, D. K. Carpenter, and S. Newman, *J. Phys. Chem.,* **62,** 1586 (1958).

(27) See, for example, T. Shimanouchi, *Pure and Applied Chem.,* **12,** 287 (1966).

(28) T. Makimoto, T. Tsuruta, and J. Furukawa, *Makromol. Chem.,* **50,** 116 (1961).

(29) T. Yoshino, Y. Kikuchi, and J. Komiyama, *J. Phys. Chem.,* **70,** 1059 (1966).

(30) P. J. Flory, J. E. Mark, and A. Abe, *J. Am. Chem. Soc.,* **88,** 639 (1966).

(31) J. E. Mark, R. A. Wessling, and R. E. Hughes, *J. Phys. Chem.,* **70,** 1895 (1966).

(32) R. A. Wessling, J. E. Mark, E. Hamori, and R. E. Hughes, *J. Phys. Chem.,* **70,** 1910 (1966).

(33) P. J. Flory and J. D. Baldeschwieler, *J. Am. Chem. Soc.,* **88,** 2873 (1966).

(34) T. M. Connor and K. A. McLauchlan, *J. Phys. Chem.,* **69,** 1888 (1965).

(35) N. Sheppard and J. J. Turner, *Proc. Roy. Soc. (London),* **A252,** 506 (1959).

Optical and NMR Studies of the α-Helix and the Helix-Coil Transition

1. *Introduction.* We turn now from vinyl polymers, where our configurational problems concern the relative placement of pseudoasymmetric centers, to polypeptide chains, where we deal with sequences of true asymmetric centers of known handedness. Here, our concern is with conformation rather than configuration. Because of the presence of asymmetric centers, additional tools are available to us which we could not employ before: optical rotatory dispersion (ORD) and the associated circular dichroism (CD). Let us consider these phenomena briefly.

2. *Circular Birefringence.* Electromagnetic radiation involves the propagation of both electric and magnetic fields. If a light beam is monochromatic and *linearly polarized*, the magnitude of the electric field at any point in a wave train will oscillate sinusoidally in a single plane with a frequency $\nu = c/\lambda$, where c is the velocity of propagation and λ the wavelength in the medium employed. The associated magnetic field will oscillate in the same manner but in a plane perpendicular to that of the electric field. Since we are primarily concerned with the interaction of the electric field with matter, we shall confine our attention to it.

It is convenient and realistic to regard the vector E representing the electric field of a linearly polarized wave train as the resultant of two equal vectors rotating in phase in opposite directions with frequency ν. If, as in Fig. IV-1, we imagine the wave train to be directed toward the observer, the clockwise-rotating vector will be designated E_R and the counterclockwise-rotating vector will be designated E_L. At any instant, E_R and E_L will always make equal angles with the axis p representing the plane of polarization. E_R taken alone would represent the behavior of the electric field of a *right-circularly polarized* wave train of frequency ν; while E_L

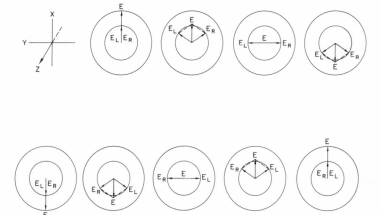

Fig. IV-1. Plane polarized light expressed as the resultant of counter-rotating vectors E_L and E_R.

would represent the electric field of a *left-circularly polarized* wave train of this frequency. Thus, a beam of linear polarized light is to be thought of as consisting of left- and right-circularly polarized beams of equal intensity and frequency. It is actually experimentally possible, by means we shall not discuss here, to resolve linear polarized light into its circular components. This is the method by which circularly polarized light is produced.

The vectors E_R and E_L will make equal angles with p only if the right- and left-circularly polarized components travel with the same speed in the medium concerned, *i.e.,* if the refractive index of the medium for the right-circularly polarized component, n_R, is the same as that for the left-circularly polarized component, n_L. It is a fundamental characteristic of optically active media that the velocities of the circular components will in general *not* be equal, *i.e.,* such media exhibit circular birefringence. If, for example, E_R travels faster than E_L, *i.e.,* $n_L > n_R$, it can be readily shown (see Fig. IV-2)

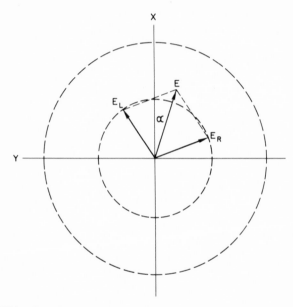

Fig. IV-2. Rotation of the plane of polarized light as a result of circular birefringence.

that, although the two components have the same ν, and continue to counter-rotate in phase, the resultant E will be tipped clockwise through an angle α. Conversely, if $n_L < n_R$, E will be rotated counterclockwise. Quantitatively, it is found that the rotation of the phase of polarization is given by

$$\alpha = \frac{1.8 \times 10^{10}}{\lambda} (n_L - n_R) \tag{IV-1}$$

where α is expressed in the traditional units of degrees rotation per decimeter of path length, and λ is in *millimicrons* (10^{-7} cm). The specific rotation is given by

$$[\alpha] = \alpha/c \qquad \text{(IV-2)}$$

where c is the concentration of the optically active substance in grams per cc of solution; the molar rotation $[m]$ is given by

$$[m] = \frac{\alpha M}{100 c} = \frac{[\alpha] M}{100} \qquad \text{(IV-3)}$$

where M is the molecular weight of the optically active substance. (The molar rotation is also commonly designated as $[\phi]$.) In reporting measurements on polymers, it is customary to express the results in terms of the molar rotation per residue or monomer unit. If more than one kind of residue is present, this quantity is termed the *mean residue rotation*. The units of $[m]$ are *degrees-cm^2-decimole.$^{-1}$*

If the refractive index of the medium is known for the wavelength employed, one may employ the Lorentz correction factor to reduce the mean residue rotation to the value corresponding to vacuum

$$[m'] = \frac{3}{(n_\lambda^2 + 2)} \frac{[\alpha] M}{100} \qquad \text{(IV-4)}$$

The quantity $[m']$ is commonly called the *reduced mean residue rotation*. It permits a more meaningful comparison of rotations observed for the same substance in different solvents.

In modern spectropolarimeters, it is common to measure values of α of the order of millidegrees. One millidegree rotation per decimeter path length corresponds at 250 mμ to a value of $n_L - n_R$ of *ca.* 1.4×10^{-11}. It can thus be readily appreciated that in the phenomenon of optical rotation, nature has provided a powerful means of amplification of the effects of circular birefringence, which itself would usually be far too small to measure directly.

Because of the sensitivity of modern instruments it is common to employ path lengths of the order of 1 mm or less

rather than the decimeter paths traditional in visual polari-
meters. A convenient expression for the calculation of molar
rotation is then

$$[m] = \frac{\alpha'}{[M]l} \qquad (IV\text{-}5)$$

where α' is the observed rotation in millidegrees, $[M]$ is the
concentration of the optically active substance in moles per
liter, and l is the path length in millimeters.

3. *Circular Dichroism.* In addition to exhibiting different
refractive indices for right- and left-circularly polarized light,
an optically active medium will also exhibit different extinction coefficients for these components of the linearly polarized beam. For the molar extinction coefficient in an optically
inactive medium we have

$$\epsilon = \frac{\log_{10}(I_0/I)}{[M]l} \qquad (IV\text{-}6)$$

where I_0 and I are the intensities of the entering beam and the
beam after passing through 1 cm of a solution of molar concentration $[M]$. For an optically active medium we have for
the molar circular dichroism

$$\Delta\epsilon = \epsilon_L - \epsilon_R = \frac{1}{[M]l}\left(\log_{10}\frac{I_0}{I_L} - \log_{10}\frac{I_0}{I_R}\right)$$

$$= \frac{\log_{10}(I_R/I_L)}{[M]l} \qquad (IV\text{-}7)$$

The significance of the differing extinction coefficients may
be seen in Fig. IV-3. The vectors E_L and E_R are now unequal
in length. Let us suppose that the left-circular component is
the more strongly absorbed. E_L is then shorter than E_R. Let
us suppose for the moment that n_L and n_R are equal, as in
Fig. IV-3a. (We shall see below that this does not necessarily
contradict our assumption of an optically active medium.)
The tip of the vector E, the resultant of E_L and E_R, now no
longer oscillates along the line p representing the plane of

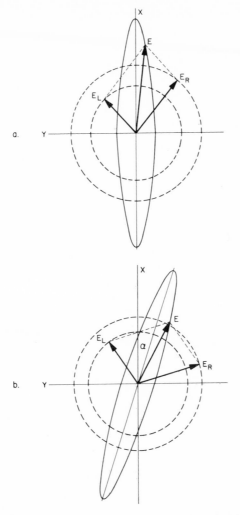

Fig. IV-3. Positive circular dichroism at (a) the Cotton band center (no rotation of the resultant E); (b) at longer wavelength than the band center (rotation of E to the right).

polarization, but instead traces out an ellipse. The semimajor axis of this ellipse is $E_L + E_R$ and coincides with p; the semiminor axis is $E_R - E_L$. Since $E_R - E_L$ is usually very small, the ellipse is a very elongated one. The light emerging from the medium, then, is actually no longer linearly polarized but weakly *elliptically polarized*. The *ellipticity* is defined as the angle whose tangent is the ratio of the minor to the major axis of the ellipse. For a 1-cm path, the *molar ellipticity* is given with sufficient accuracy by

$$[\theta] \cong 2.303\left(\frac{4500}{\pi}\right)(\epsilon_L - \epsilon_R) \qquad (IV\text{-}8a)$$

$$\cong 3300(\epsilon_L - \epsilon_R) \qquad (IV\text{-}8b)$$

Many authors report the molar circular dichroism itself, since this is the quantity directly provided by the circular dichroism spectrometer. Equations IV-8 give the relationship of these quantities.

4. *Molecular Origin of Optical Activity.* The quantum mechanical basis of optical activity was given by Rosenfeld[1] many years ago, and since that time it has been well understood that to rotate the plane of the polarized light and to exhibit circular dichroism, a molecule must develop, in the course of the electron oscillations induced by the impinging wave train, both an electric moment (leading to absorption) and a *magnetic moment* which has a finite component in the direction of the electric moment. Mechanistically, one can picture the electron as proceeding in a twisting path as it oscillates, thus acting like current in a solenoid and generating a magnetic moment. It can be shown that such twisting is possible only in a molecule which cannot be superimposed on its mirror image; the correlation of optical activity with such structural asymmetry has of course long been known to the organic chemist.

The strength of the rotation and circular dichroism of a particular electronic transition will be proportional to the product of the electric and magnetic moments associated with the transition. Various type of models have been employed in an attempt to calculate the magnitudes of these moments, including those which require the coupling of the oscillations of electrons in different parts of the molecule,[2,3] and those which envision single-electron jumps in the molecule.[4] We shall not describe these further here. The reader is referred to references 5 and 6 for a fuller discussion.

It is possible for a transition to have a weak electric moment and a strong magnetic moment, *i.e.,* to be weak in absorption but strong in rotation and circular dichroism. Such transitions are referred to as "magnetic dipole allowed"; a transition strong in absorption but weak in rotation and circular dichroism is referred to as "electric dipole allowed."

5. *The Dependence of Circular Birefringence and Circular Dichroism on Wavelength: Cotton Bands.* It is to be expected that both circular birefringence and circular dichroism will depend strongly upon wavelength. From the classical viewpoint (which can carry us quite far in discussing these phenomena), one may regard the response of the molecular electrons to the impinging field as analogous to the forced oscillations of a mechanical system. It has long been realized that as the resonant frequency of the system is approached and traversed, it will exhibit responses in-phase (or *dispersive*) and out-of-phase (or *absorptive*). This duality is very general in nature. Thus, the variation of refractive index (irrespective of whether the system is optically active or not) with wavelength is a dispersive phenomenon. Its change through the band is usually relatively small, but depends upon the intensity of absorption. Typical behavior is shown in Fig. IV-4. At the center of the absorption band, n passes through a value equal to that which it approaches asymptotically at wavelengths far removed from the band (assuming there to be but one band).

A plot identical in form to Fig. IV-4 describes the dependence of circular dichroism and optical rotation upon wavelength in the region of an absorption band (Fig. IV-5). As is well known, an absorption band characterized by optical ac-

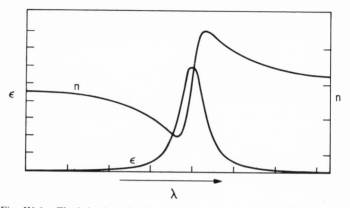

Fig. IV-4. The behavior of the refractive index of a medium in the neighborhood of an absorption band.

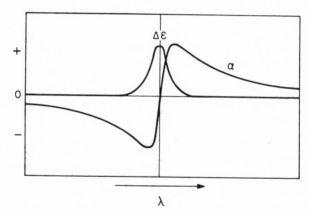

Fig. IV-5. A Cotton band as observed (a) by optical rotatory dispersion and (b) by circular dichroism.

tivity (a term which will be used to include both circular dichroism and optical rotation) is termed a *Cotton band*,* after its first discoverer.[7] It is a fundamental property of dispersion bands that they have zero intensity at the band

---

*In the literature, this term is not always used with respect to both circular dichroism and optical rotation, but there seems little point in this distinction.

center but extend with measurable intensity far out in both directions from the center. Thus, rotations have been traditionally measured at the wavelength of the D line of sodium (589 mμ), which is over 400 mμ removed from many important chromophores, as for example in carbohydrates. The circular dichroism spectrum, on the other hand, being a miniature reproduction of the absorption band, can be observed only in its immediate neighborhood. The Cotton band shown in Fig. IV-5 is *positive*: the rotation is positive on the long wavelength side, and $\epsilon_L - \epsilon_R$ is positive at all points. For a negative Cotton band, these curves would be inverted.

In earlier work, instrumental limitations made it impossible to observe directly the Cotton bands in most polypeptide and protein spectra, because, in the absence of chromophores such as heme units, the bands of principal interest are in the ultraviolet; those of the polypeptide backbone itself are in the far ultraviolet (220–185 mμ) and very probably in the vacuum ultraviolet region (below 185 mμ) as well. This latter region has not yet been explored. Under these circumstances, recourse was had to so-called Drude equations, either of the one-term type (Eq. IV-9) or two-term type (Eq. IV-10)

$$[m]_\lambda = \frac{a_c \lambda_c^2}{\lambda^2 - \lambda_c^2} \qquad \text{(IV-9)}$$

$$[m]_\lambda = \frac{a_0 \lambda_0^2}{\lambda^2 - \lambda_0^2} + \frac{b_0 \lambda_0^4}{(\lambda^2 - \lambda_0^2)^2} \qquad \text{(IV-10)}$$

These equations describe the rotation in the visible and near ultraviolet as a function of wavelength. A one-term Drude plot describes the dispersion adequately if there is only one predominant Cotton band, particularly if it is in the far ultraviolet, which, as we shall see, is characteristic of polypeptides in the random-coil form. For polypeptides in the α-helical form (see below), this equation cannot be made to fit the data, and Eq. IV-10 must be used. Here λ_c, a_0, and b_0 are empirical constants. The value of a_0 is positive for right-handed α-helices and depends rather strongly on solvent; it is generally in the range of $+170 \pm 50$; b_0 and λ_0 are independent of solvent and type of side-chain, and are close to -630 and 212 mμ, respectively, for right-handed α-helices. For left-handed α-helices, b_0 is of similar magnitude but positive. (For summaries see references 5, 8, and 9.)

Although much valuable work has been accomplished using this approach, it has become outmoded by the development of ORD and CD spectrometers capable of measurement down to 185 mμ. We shall make no further reference to Drude-plot data in our subsequent discussion.

6. *Band Overlap in ORD and CD Spectra.* The fact that ORD and CD bands may be of either sign leads to certain complications in spectra having overlapping bands of both signs. These are particularly evident in CD spectra, as illustrated in Fig. IV-6. Wellman *et al.*[10] have pointed out that

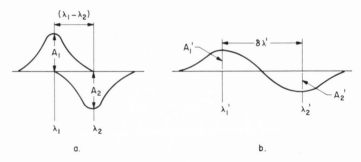

Fig. IV-6. CD band overlap: (*a*) True band positions and intensities; (*b*) Experimentally observed bands.

when there are two equal, oppositely signed, overlapping CD bands [(*a*) in Fig. IV-6], their summation, *i.e.,* the experimentally observed spectrum, will show them as weaker $(A_1', A_2' < A_1, A_2)$ than they actually are (which is fairly obvious), but will also cause them to appear further apart than their true separation [(*b*) in Fig. IV-6]. It can be easily shown that for Gaussian bands, the true separation of which $(\lambda_1 - \lambda_2)$ is small compared to their width $\Delta\nu$ (not true in Fig. IV-6), the *apparent* separation $\delta\lambda'$ will never be less than

$\sqrt{2}\Delta\nu$. Correct intensities and positions for such bands can be obtained by analog- or digital-computer simulation of the observed spectrum.[11]

7. *The α-Helix.* This chapter will center chiefly on the α-helix and its conformational changes, as observed by both high-resolution NMR and circular dichroism. In particular, we shall be concerned with the α-helix–random coil transition. This phenomenon has, of course, been intensively studied both experimentally and theoretically (see refs. 5, 8, and 9) but there nevertheless are some unresolved problems. In particular, the question of why the transition occurs at all appears to be quite unanswered for organic solvent systems. We may not be able to give a complete answer here either, but at least will furnish some additional evidence.

In Fig. IV-7 is shown the now very familiar structure of the α-helix.[12] The structure represented is that of poly-L-alanine in the right-handed helical conformation. This representation claims two somewhat novel features: first, it is a pair of stereoscopic views; second, it was not drawn by a draftsman, but was generated directly on microfilm by the General Electric 635 computer, using a program adapted for this purpose by R. L. Kornegay (Bell Telephone Laboratories) from a more general program devised by C. K. Johnson (Oak Ridge National Laboratories).

As it occurs in nature and in many synthetic polypeptides, the α-helix has approximately 18-fold symmetry, *i.e.,* it repeats exactly every 18 amino acid residues. There are 3.6 residues per turn and a residue translation of 1.49 Å, *i.e.,* as we pass from a point in one residue to the corresponding point in the next, we move 1.49 Å along the helical axis.

In Fig. IV-8 are shown the absorption spectrum and the circular dichroism spectrum characteristic of a polypeptide in the α-helical conformation. These spectra are of poly-γ-methyl-L-glutamate (DP 400) in hexafluoroisopropanol solution. They are in close agreement with those presented by Holzwarth and Doty[13] for this polypeptide in trifluoro-ethanol, but the CD band intensities are slightly smaller than they report. In Fig. IV-9, the CD spectrum has been decom-

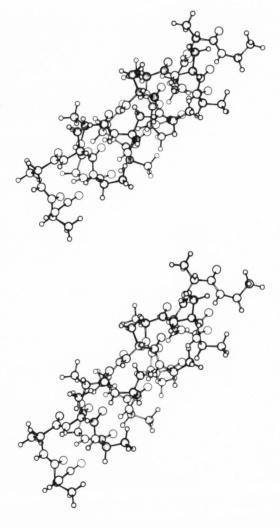

Fig. IV-7. Stereo views of α-helix.

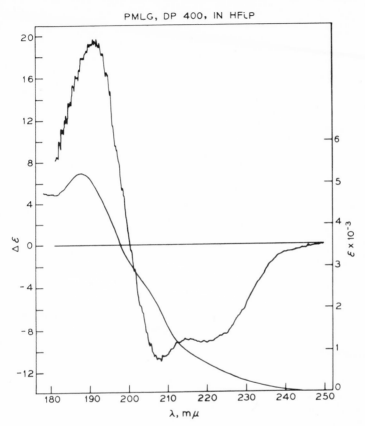

Fig. IV-8. Absorption and CD spectra of poly-γ-methyl-L-glutamate, 2×10^{-2} M in hexafluoroisopropanol.

posed by an analog computer (Sec. 6) into the constituent bands as now recognized, and the relative intensity of each is indicated.

The presently accepted interpretation of these CD bands, and of the corresponding bands of ORD and absorption spectra, is based in part on the theoretical studies of Moffitt,[14,15] who first predicted that strong rotatory (and CD) bands of opposite sign should be observed for regular arrays

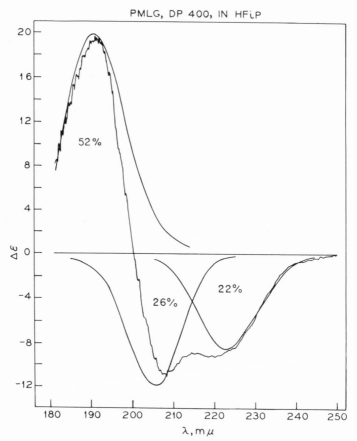

Fig. IV-9. The CD spectrum of Fig. IV-8 decomposed into n-π^*, $\parallel \pi$-π^*, and $\perp \pi$-π^* bands.

of interacting chromophores because of resonance coupling of their electric dipole transition moments. It was further predicted that the band at longer wavelength would be polarized parallel to the helical axis and that the band at shorter wavelength would be polarized perpendicularly to the helical axis. These predictions have been amply borne out by a host of experimental studies, beginning with those of Rosenheck and

Doty[16] and of Tinoco *et al.*[17] (For a review see references 5, 8, 9, and particularly reference 13.) The interpretation of the spectra in Figs. IV-8 and IV-9 is as follows:

  a. At 190 mμ and 205 mμ are the oppositely signed, exciton-split π-π* bands of the peptide chromophore. The positive band is polarized perpendicularly to the helical axis; the negative band is polarized parallel to the helical axis. The positive band has twice the intensity of the negative band. (This last observation is not in accord with Moffitt, who predicted equal intensities.)
  b. At 223 mμ is the negative n-π* peptide band, quite strong in dichroism, but weak in absorption. This transition was not discussed by Moffitt; its theoretical interpretation has been considered by Schellman and Oriel[18] and by Tinoco and co-workers,[19] among others.

We are concerned also with the left-handed α-helical conformation which poly-L-aspartate chains are believed to prefer. There appear to be few if any published circular dichroism data for these polymers, the left-handed configuration being deduced from b_0 values obtained from optical rotatory dispersion measurements.[20] From our own limited data, we know that poly-β-methyl-L-aspartate and poly-β-benzyl-L-aspartate give *positive n-π*** CD bands in about the same position as the negative n-π* band of a typical right-handed α-helix.

When transition from the right-handed α-helix to the random coil occurs in aqueous solution, it has been reported[11,13] that the negative n-π* band gives place to a very weak positive band. (As we shall see, this is not necessarily the case in organic solvents, where a weak negative n-π* band apparently persists in the random-coil conformation.) The π-π* exciton-split bands are replaced by a single negative band at about 200 mμ. In the work to be described, the π-π* transition could not be observed because of high absorption, but the n-π* band furnished a satisfactory measure of the helix-coil transition.

8. *NMR Observations of the α-Helix–Random Coil Transition.* The first NMR study of the helix-coil transition was reported in 1959 by Bovey and Tiers,[21] who observed an extreme broadening of the NMR peaks of poly-γ-benzyl-L-glutamate (PBLG) in trichloroethylene, but were able to see reasonably well-resolved spectra for both helix and coil on addition of appropriate quantities of trifluoroacetic acid (TFA). Several other authors have since examined this and closely related systems.[22-30] More recently, studies of poly-γ-benzyl-L-glutamate and poly-β-benzyl-L-aspartate have been carried out at 220 MHz.[31] In our work, deuterochloroform has been used as the helix-supporting solvent and TFA as the helix-breaking solvent. For both studies we have used polymers of low molecular weight (DP *ca.* 50) and of high molecular weight (DP *ca.* 1000).

In Fig. IV-10 are shown 100 MHz spectra of PBLG of DP 55 in chloroform alone, with 15% TFA, and with 30% TFA. These spectra were run at 50° and are representative of the many that were run. All polymer solutions contained 10% (w/v) of polymer. There is a marked broadening of all peaks in $CDCl_3$. The peaks of the backbone protons, NH and α-CH, are so broadened that they seem to have disappeared, as has been previously noted by Goodman and Masuda.[22] However, on increasing the spectrometer gain, they can be readily seen. The protons of the side-chain are progressively less broadened as we move out from the glutamyl methylenes (β-CH_2 at *ca.* 7.7τ and γ-CH_2 at *ca.* 7.5τ) to the benzylic methylene group (4.96τ) and to the phenyl group (2.76τ). Upon addition of 15% of trifluoroacetic acid (as little as 2.5% has a nearly equal effect), all peaks narrow considerably, and the NH and α-CH resonances become clearly visible. However, their positions are the same and the chains must be still largely helical, for the circular dichroism measurements, which we shall discuss below, show an n-π* band at 224 mμ with an intensity, Δε, of −6.4 l-mole^{-1}-cm^{-1}. At 30% TFA (*c*), all peaks narrow further, but the effect is not striking. Circular dichroism measurements show that the polymer is now a random coil, the n-π* band having nearly disappeared.

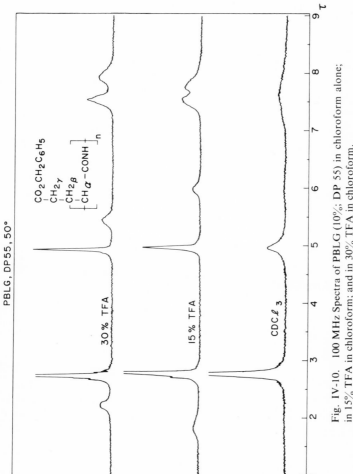

Fig. IV-10.   100 MHz Spectra of PBLG (10%; DP 55) in chloroform alone; in 15% TFA in chloroform; and in 30% TFA in chloroform.

In Fig. IV-11 are shown corresponding spectra for the PBLG of DP 1000. In pure chloroform, the entire spectrum appears to have vanished completely when recorded over the usual 1000 cps sweep width. The smaller inset spectrum, recorded over a 3000 cps sweep width, shows that it is actually still there (as of course it must be) but very greatly broadened. The only visible peak is the phenyl resonance, now about 250 cps in width. Measurable intensity is spread out over about 4000 cps, as integration of the spectrum clearly shows. In the presence of 2.5% TFA, the peaks are likewise extremely broad. At 15% TFA the polymer is still helical, but the peaks are very much narrower (probably by a factor of at least 100); they are, however, markedly broader than for the DP 55 polymer under the same conditions. The peaks of the random coil exhibit the same width as those of the low-molecular-weight polymer in the random-coil state; the spectra are indistinguishable.

As previously pointed out,[21] the extreme broadening in deuterochloroform is very likely due to aggregation. The existence of PBLG in a liquid crystalline state under these conditions has been established.[32] The broadening is closely analogous to that exhibited by native proteins, but somewhat more extreme. It is strongly dependent upon molecular weight. Upon adding TFA, these aggregates are broken up. A considerable dependence of the linewidth of the free helices upon molecular weight is still noticeable, as might be expected for a rod-like macromolecule, but it is much less marked. In the random-coil state, there is no dependence upon molecular weight, since now local segmental motion determines the line-width. This last is the behavior normally characteristic of vinyl polymers in solution.

Figures IV-12 and -13 show similar data for poly-β-benzyl-L-aspartate of low and high molecular weight. The spectral broadening in chloroform is comparable, but is not quite so great: both the phenyl and benzyl protons can now be discriminated. The onset of narrowing occurs at much lower acid concentration for both the low- and high-molecular-

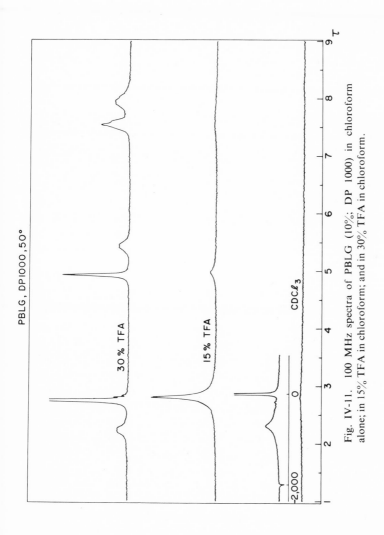

Fig. IV-11. 100 MHz spectra of PBLG (10%; DP 1000) in chloroform alone; in 15% TFA in chloroform; and in 30% TFA in chloroform.

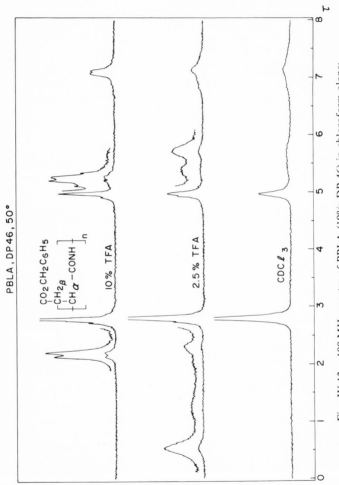

Fig. IV-12. 100 MHz spectra of PBLA (10%; DP 46) in chloroform alone; in 2.5% TFA in chloroform; and in 10% TFA in chloroform.

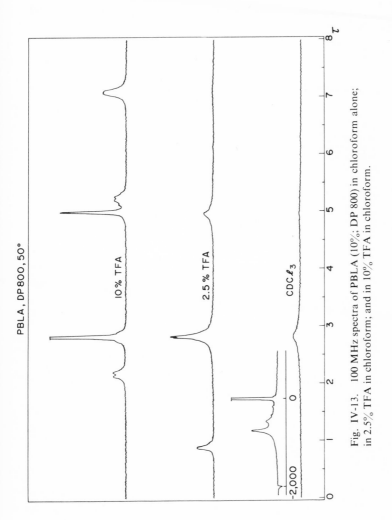

Fig. IV-13. 100 MHz spectra of PBLA (10%; DP 800) in chloroform alone; in 2.5% TFA in chloroform; and in 10% TFA in chloroform.

weight polymer. In 30% TFA, the NH resonance is a doublet and the α-CH a binomial quartet, indicating approximately equal couplings (*ca.* 7 cps) of the α-CH to the β-CH and NH protons. This clearly is a measure of the average local conformation of the random coil, but unfortunately we do not at present know the dependence of the vicinal coupling upon the H-N-C-H dihedral angle.

Let us turn now to a consideration of the chemical shifts

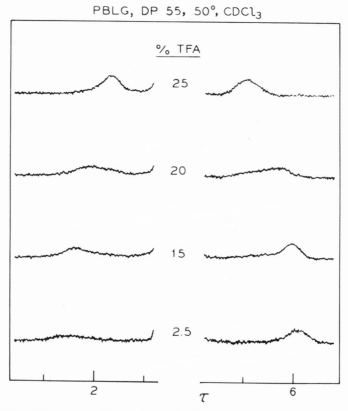

PBLG, DP 55, 50°, CDCl$_3$

% TFA

25

20

15

2.5

2

τ

6

Fig. IV-14.   NH and CH peaks of PBLG (10%; DP 55) as a function of TFA concentration in chloroform.

of the various polypeptide protons as a function of conformation. Certain obvious trends can be seen in Figs. IV-10–13. The behavior of the NH and α-CH protons is shown in greater detail in Figs. IV-14 and -15, which represent the low-molecular-weight polymers at 50°. For PBLG, the α-CH peak remains unchanged at 6.00τ until about 20% TFA, when it begins an *apparent* downfield shift, accompanied by a noticeable broadening. At 25% TFA, this apparent shift is com-

PBLA, DP 46, 50°, CDCl$_3$

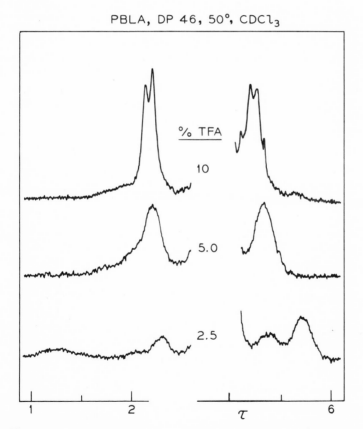

Fig. IV-15.   NH and α-CH peaks of PBLA (10%; DP 46) as a function of TFA concentration in CHCl$_3$.

plete, the α-CH peak appearing at *ca.* 5.4τ. At the same time, the NH protons become markedly *more* shielded. These changes presumably correspond to the transition from α-helix to random coil. Closer examination of these spectra suggests the conclusion that instead of a shift, there is actually a decrease of the helical peak accompanied by a growth of the coil peak and a broadening of both. We shall consider this point, which is a significant one, a little later.

These same features are exhibited in the poly-β-benzyl-L-aspartate spectra, shown in Fig. IV-15. The NH peak "moves" even more markedly upfield when the transition to the random coil occurs; this is complete at much lower TFA concentration—about 5%—than in the glutamate system. We observe that even in deuterochloroform alone (not shown in Fig. 15), there is appreciable intensity in the random coil α-CH and NH peak positions. These probably correspond to chains too short to form a helix.

The appearance of two α-CH and NH peaks might be expected in the transition region if the rate of equilibration of the helix and coil were smaller than the separation of their resonance peaks, expressed in cps. Ferretti[28] has observed a doubling of the NH and α-CH peaks for poly-β-methyl-L-aspartate and poly-L-leucine in the region of their transitions, and has concluded that the life-times of the two forms are probably at least 10^{-2} sec. (The molecular weights of the polymers were not stated.) Similar conclusions have been drawn by Bradbury *et al.*[26,29] for some polyamino acids, *e.g.*, poly-D-norleucine, but not for all. In assessing this conclusion, it must be remembered that temperature-jump measurements have shown[33] that for polyglutamic acid in aqueous solution the lifetimes are less than 10 microsec. It therefore appears that in organic solvents the rate of equilibration is 10^3 to 10^4 times slower than in water.

In order to test further this somewhat surprising result, it is desirable to observe polymers of high molecular weight, having no substantial fraction of very short chains. However, such polymers, when observed at 100 MHz (Figs. IV-11 and

-13), or even at 220 MHz, give peaks so broad that one cannot hope to see separate helix and coil resonances—at least not in PBLG and PBLA. We have therefore reexamined PBLG of DP 50 at 220 MHz. Figure IV-16 shows the spectrum of this polymer at 35° in CDCl₃ containing 16% (v/v) of

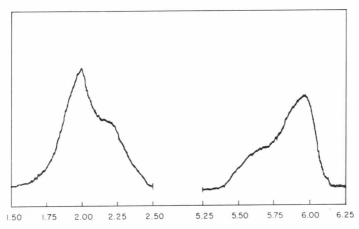

Fig. IV-16. α-CH resonance of PBLG (10%; DP 55) observed at 220 MHz; 16% (v/v) TFA in CDCl₃ at 35°.

TFA. At lower temperatures it is observed that the helix is less stable and requires a lower concentration of TFA to disrupt it. At 35° at this concentration we are about one-third of the way through the transition from helix to random coil, and it is very clear (as it was not at 100 MHz) that two NH and α-CH peaks are present. Ferretti's conclusions appear to be strengthened. The system does not give a single, averaged peak which moves downfield. Rather, the downfield (coil) peak grows at the expense of the other.

9. *Correlation of NMR and CD Observations.* It might be asked whether the observed changes in the NMR spectra are actually due to the helix-coil transition itself or perhaps to a chemical reaction such as the protonation of the peptide oxy-

gen atoms:

$$HA + \underset{/}{\overset{O}{\diagdown}}C-N\underset{\diagdown H}{\diagup} \rightleftharpoons A^- + \underset{/}{\overset{HO}{\diagdown}}C{=}N\underset{\diagdown H}{\diagup}$$

Such protonation is known to occur for small molecule amides, at least if the acid is strong enough.[34-36] This might precede but not necessarily coincide with the actual transition. To answer this question, let us look a little more closely at the circular dichroism measurements. Those shown in Fig. IV-17 were made at 33°, using the same solutions used for the 100 MHz NMR measurements. Because of solvent and polymer absorption in these concentrated solutions, it was not possible to make measurements beyond 222 mμ, but this suffices to reach the n-π^* extrema for both the polyglutamate and polyaspartate systems. In Fig. IV-18, the α-CH peak positions and $\Delta\epsilon$ are plotted versus the acid concentration. Both appear to depend on the acid concentration in the same way, and, therefore, one may logically (but not rigorously) conclude that both reflect the helix-coil transition. Very similar observations (but not including the NMR data) have been independently reported by Quadrifoglio and Urry.[37]

10. *Origin of the Helix-Coil Transition.* These experiments raise once again a knotty and still unanswered question: just what is it that causes the transition from helix to coil to occur in these systems as the concentration of acid is increased? The marked upfield shift of the TFA carboxyl protons caused by the polypeptide (reported by Stewart *et al.*[24] and confirmed in our studies) clearly points to a strong interaction of some kind. Stewart *et al.*[24] also studied the behavior of small model molecules, *N*-methylacetamide and *N,N*-dimethylacetamide, in the chloroform-TFA system and found a very marked *deshielding* of the TFA protons.[38] They believed this pointed to the formation of an ion pair of protonated amide and trifluoroacetate ion. They further concluded that TFA does *not* protonate the polypeptide amide oxygens, since the change in carboxyl peak position is the op-

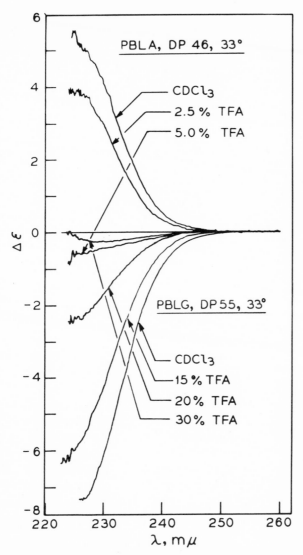

Fig. IV-17. CD spectra of 10% solutions of PBLG (DP 55) and PBLA (DP 46) as a function of TFA concentration in CHCl₃.

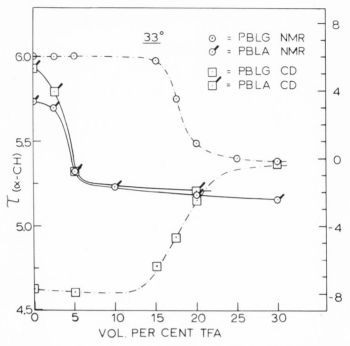

Fig. IV-18. $\Delta\epsilon$ and $\tau_{(\alpha\text{-CH})}$ of 10% solutions of PBLG (DP 55) and PBLA (DP 46) plotted *versus* TFA concentration in CHCl$_3$; 33°.

posite direction. There is, however, infrared evidence for such polypeptide protonation. Klotz and Hanlon and co-workers[39-41] have observed band changes in the NH overtone region which have been interpreted as indicating that proton-ation can occur for both model compounds and polymers, and they conclude that protonation causes the helix-coil transition. However, there are grounds for doubting that tri-fluoroacetic acid is strong enough to cause protonation of either model amides or polypeptide chains.

This doubt has been generated by a model compound ex-periment. The requirements of a suitable model amide are that it should have a strong n-π^* band, but should be rigid and incapable of conformational change. Small molecule amides giving observable n-π^* CD bands are very scarce; most simple amides show no observable dichroism in this region of the spectrum. Litman and Schellman[42] have observed a weak n-π^* CD band in L-3-aminopyrrolidone, and weak n-π^* bands have been found by us and by Balasubramanian and Wetlaufer[43] in a number of diketopiperazines. These latter compounds are subject, however, to possible conformational changes. A more suitable compound is *d*-oxolupanine

which was observed some time ago by Dr. J. W. Longworth, then in our laboratory, to have a strong ORD Cotton band in the n-π^* region. Its conformation appears from molecular models to be rigid. The CD spectra of this diamide in the region of major interest are shown in Fig. IV-19 in a number of solvents. In chloroform, a band is seen at 230 mμ, comparable in intensity ($\Delta\epsilon = 5.4$) to that of poly-β-benzyl-L-aspartate and remarkably strong for a nonpolymeric molecule. The CD spectrum is trifluoroethanol is similar, and even somewhat stronger ($\Delta\epsilon = 7.2$), but shifted to 221 mμ. In addition to the positive band, there is a negative band (not shown) at 202 mμ, in the same position as the absorption maximum. It seems reasonable to assign these to n-π^* and π-π^* transitions, respectively, just as in the α-helix spectrum. If protonation were to occur to a substantial extent, the n-π^* band should decrease in intensity or disappear, for if the representation of the protonated form on p. 124 is correct, the nonbonded electrons of the unprotonated carbonyl oxygen are now bonded. In concentrated sulfuric acid, this band does indeed disappear, but it reappears on dilution with water, showing that the compound has not been decomposed.

D-OXOLUPANINE

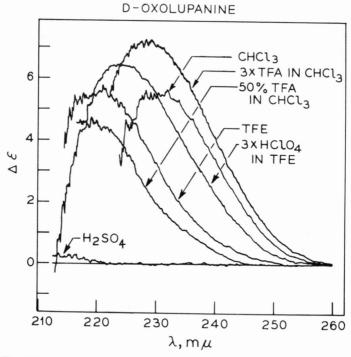

Fig. IV-19. The n-π^* region of the CD spectrum of d-oxolupanine in CHCl$_3$ and in CHCl$_3$ with added TFA; in TFE and in TFA with added HClO$_4$; and in concentrated H$_2$SO$_4$.

(The π-π^* band and the absorption maximum shift in position, and the former changes sign.) But in the presence of a three-fold and 600-fold molar excess of trifluoroacetic acid, the band does not disappear and even increases slightly in intensity. It appears that the amide groups in oxolupanine should be at least as strongly basic as those of a polypeptide chain, and probably more so, since they are N,N-dialkyl amide groups, which are normally stronger than N-mono-alkyl amides. The carbonyl groups do not appear to be any more sterically hindered in the model than in the polypeptide.

We therefore conclude that the helix-coil transitions reported so far for non-aqueous solvents do not involve proton loss and gain by the polymer chain, and that other interpretations must be sought for those experimental results which have been interpreted in this way. We thus must appeal in part to the familiar hydrogen-bond competition as being responsible for the transition:

$$HA \cdots HA \; + \; -C{=}O \cdots HN \; \rightleftharpoons \; -C{=}O \cdots HA \; + \; HA \cdots HN$$

We assume that this equilibrium runs to the right when HA is reasonably strong, and that increasing temperature pushes it to the left for the poly-γ-benzyl glutamate system, and probably for most right-handed α-helices. It further appears that for the poly-β-benzyl-L-aspartate system, the position of this equilibrium is unaffected by temperature.

Recent potential-energy calculations by a number of authors (see Chapter V) have shown that α-helical conformations tend to be preferred even in the *absence* of intramolecular hydrogen bonding, as a result of van der Waals interactions, torsional potentials and peptide dipole-dipole interactions. Therefore, the disruption of hydrogen bonding is itself not really a sufficient answer. It must be that there is in addition a substantial difference in solvation energy between the helix and the coil aside from that involving hydrogen bonding, and that more refined energy calculations should take this into account. Also, we must of course recall that, other things being equal, there is a substantial positive entropy term which encourages random-coil formation whatever the heat terms involved may be. It is to be expected, however, that the magnitude, and in some cases even the sign, of this term will also be strongly dependent on the nature of the solvent.

REFERENCES FOR CHAPTER IV

(1) L. Rosenfeld, *Z. Physik*, **52**, 161 (1928).
(2) W. Kuhn, "Stereochemie," Deuticke, Leipzig, 1933, p. 317.
(3) J. G. Kirkwood, *J. Chem. Phys.*, **5**, 479 (1937).
(4) E. V. Condon, W. Altar, and H. Eyring, *J. Chem. Phys.*, **5**, 753 (1937).
(5) J. A. Schellman and C. Schellman, in "The Proteins," ed. by H. Neurath, Academic Press, New York, 1964, Vol. II, pp. 1–128.

(6) A. Moscowitz, in "Optical Rotatory Dispersion" by C. Djerassi, McGraw-Hill Book Co., New York, 1959, Chap. 12.

(7) A. Cotton, *Ann. Chim. Phys.*, [vii] **8**, 347 (1896).

(8) P. J. Urnes and P. Doty, *Advances in Protein Chemistry*, **16**, 401 (1961).

(9) J. T. Yang in "Newer Methods of Polymer Characterization," ed. by B. Ke, Interscience Publishers, New York, 1964, pp. 103–153.

(10) K. M. Wellman, P. H. A. Laur, W. S. Briggs, A. Moscowitz, and C. Djerassi, *J. Am. Chem. Soc.*, **87**, 66 (1965).

(11) J. P. Carver, E. Schechter, and E. R. Blout, *J. Am. Chem. Soc.*, **88**, 2550, 2562 (1966).

(12) L. Pauling, R. B. Corey, and H. R. Branson, *Proc. Nat. Acad. Sci. U.S.*, **37**, 205 (1951).

(13) G. Holzwarth and P. Doty, *J. Am. Chem. Soc.*, **87**, 218 (1965).

(14) W. Moffitt, *J. Chem. Phys.*, **25**, 467 (1956).

(15) W. Moffitt, *Proc. Nat. Acad. Sci. U.S.*, **42**, 736 (1956).

(16) K. Rosenheck and P. Doty, *Proc. Nat. Acad. Sci. U.S.*, **47**, 1775 (1961).

(17) I. Tinoco, A. Halpern, and W. T. Simpson, in "Polyamino Acids, Polypeptides, and Proteins," ed. by M. Stahmann, Univ. of Wisconsin Press, Madison, Wis., 1962, p. 147.

(18) J. A. Schellman and P. Oriel, *J. Chem. Phys.*, **37**, 2114 (1962).

(19) I. Tinoco, Jr., R. W. Woody, and D. F. Bradley, *J. Chem. Phys.*, **38**, 1317 (1963).

(20) R. H. Karlson, K. S. Norland, G. D. Fasman, and E. R. Blout, *J. Am. Chem. Soc.*, **82**, 2268 (1960).

(21) F. A. Bovey, G. V. D. Tiers, and G. Filipovich, *J. Polymer Sci.*, **38**, 73 (1959).

(22) M. Goodman and Y. Masuda, *Biopolymers*, **2**, 107 (1964).

(23) D. I. Marlborough, K. G. Orrell, and H. N. Rydon, *Chem. Comm.*, **1965**, 518.

(24) W. E. Stewart, L. Mandelkern, and R. E. Glick, *Biochemistry*, **6**, 143 (1967).

(25) K.-J. Liu, J. S. Lignowski, and R. Ullman, *Biopolymers*, **5**, 375 (1967).

(26) E. M. Bradbury, C. Crane-Robinson, and H. W. E. Rattle, *Nature*, **216**, 862 (1967).

(27) J. L. Markley, D. H. Meadows, and O. Jardetzky, *J. Mol. Biol.*, **27**, 25 (1967).

(28) J. A. Ferretti, *Chem. Comm.*, **1967**, 1030.

(29) E. M. Bradbury, C. Crane-Robinson, H. Goldman, and H. W. E. Rattle, *Nature*, **217**, 812 (1968).

(30) F. A. Bovey, *Pure and Applied Chemistry*, **16**, 417 (1968).

(31) F. A. Bovey and J. J. Ryan, unpublished observations.

(32) C. Robinson, J. C. Ward, and R. B. Beevers, *Disc. Faraday Soc.*, **25**, 29 (1958).

(33) R. Lumry, R. Legare, and W. G. Miller, *Biopolymers*, **2**, 489 (1964).

(34) G. Fraenkel and C. Niemann, *Proc. Nat. Acad. Sci. U.S.*, **44**, 688 (1958).

(35) G. Fraenkel and C. Franconi, *J. Am. Chem. Soc.*, **81**, 62 (1959).

(36) R. J. Gillespie and T. Burchall, *Can J. Chem.*, **41**, 148 (1963).

(37) F. Quadrifoglio and D. W. Urry, *J. Phys. Chem.*, **71**, 2364 (1967).

(38) W. E. Stewart, L. Mandelkern and R. E. Glick, *Biochemistry*, **6**, 150 (1967).

(39) S. Hanlon, S. F. Russo, and I. M. Klotz, *J. Am. Chem. Soc.*, **85**, 2024 (1963).

(40) I. M. Klotz, S. F. Russo, S. Hanlon, and M. A. Stake, *J. Am. Chem. Soc.*, **86**, 4774 (1964).

(41) S. Hanlon, *Biochemistry*, **5**, 2049 (1966).

(42) B. J. Litman and J. A. Schellman, *J. Phys. Chem.*, **69**, 978 (1965).

(43) D. Balasubramanian and D. B. Wetlaufer, *J. Am. Chem. Soc.*, **88**, 3449 (1966).

CHAPTER V

The Conformations of N-Disubstituted Polypeptide Chains

1. *Poly-L-proline.* In this chapter, we shall discuss the conformations of polypeptide chains in which the α-amino nitrogen atom is disubstituted and which therefore have no amide hydrogen available for hydrogen bonding. If such chains are capable of maintaining fixed helical structures, they must do so under stabilizing influences other than those provided by hydrogen bonds.

Of the polypeptides of this class, the most carefully studied is poly-L-proline. The prolyl unit is capable of existing in both *cis* and *trans* conformations:

cis   *trans*

These evidently differ in energy by only a few hundred calories, but, to judge from the NMR results on dialkylamides,[1] are probably separated by a barrier of *ca.* 20 kcal, arising from the partial double-bond character of the C—N bond.

Poly-L-proline is known to exist in two distinct forms in both the solid state[2-4] and in solution.[5-10] (For reviews, see Refs. 11 and 12.) In water and organic acids, it exists in a levo-rotatory form, designated as **II**. Blout *et al.*[8] report a strong negative band centered at 203 mμ ($[m']_{216}$ = *ca.* − 32000°) for this form in water. As normally prepared by polymerization of the *N*-carboxy anhydride and precipitation from pyridine, poly-L-proline exists in a different form, called **I**, which is dextrorotatory at 589 mμ and is stable in aliphatic alcohols[7,10] but unstable in water and organic acids. Upon dissolving in these latter solvents, it mutarotates over a period of several hours giving form **II**.[5,6,7] We shall consider this process in more detail a little later.

The structures of solid-state forms corresponding to **I** and **II** have been established by X-ray diffraction measurements on the crystalline polymer. It is found[4] that **I** is a right-handed helix with a residue translation of 1.85 Å and with the peptide bonds in the *cis* conformation. Stereo views of this structure are shown in Fig. V-1. It has approximately 10-fold

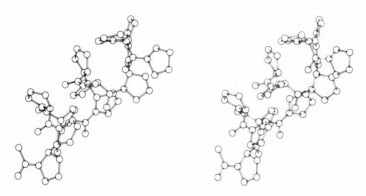

Fig. V-1.   Stereo views of the helix of poly-L-proline **I**.

symmetry. Steric restrictions seem quite severe in molecular models of this structure, which is almost as tightly wound as the α-helix. It appears to exist in solution only in relatively poor solvents, and it has been suggested[13] that its stability

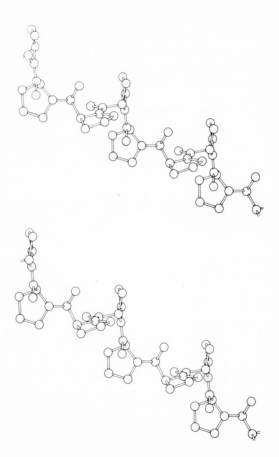

Fig. V-2.   Stereo views of the helix of poly-L-proline **II**.

can be explained only on the assumption that the groups forced into close contact interact attractively, supplanting less favorable polymer-solvent interactions.

Form **II**[2,3] is a left-handed helix with the much greater residue translation of 3.12 Å (nearly double that of the α-helix and 87% of that of the fully extended planar zigzag) and with the peptide bonds in the *trans* conformation. Stereo views of this helix are shown in Fig. V-2. It has 3-fold symmetry. The mutarotation in solution thus very probably corresponds to the *cis-trans* isomerization of the peptide bonds, which is its rate-determining step.

The mutarotation is observable at 589 mμ, but observations at a single wavelength are relatively uninformative. Much more revealing are the ORD[14] and CD[15] spectra, taken as a function of time. In Fig. V-3 is shown a selection of data for

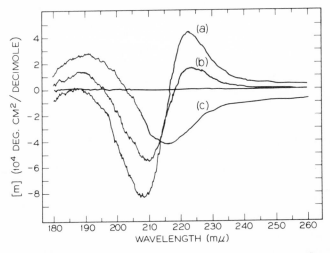

Fig. V-3. ORD spectra of poly-L-proline, 2.05 × 10^{-2} *M* in trifluoroethanol, *ca.* 30°. Curves *a*, *b*, and *c* are run at successive times. Because the rates of scanning and of mutarotation are comparable, curves *a* and *b* show some kinetic distortion. For these, the three intervals given below represent the times (minutes after initial dissolution of polymer) at which the first maximum, first minimum, and second maximum were observed; (*a*) 8, 11, 22 min; (*b*) 64, 68, 80 min; (*c*) 20 hr.

poly-L-proline (DP 140) in trifluoroethanol at 30°. It is observed that initially the polymer exhibits a strong positive Cotton band, centered at about 217 mμ. (Similar results have been observed for a poly-L-proline film grown from a catalytic surface.[16]) This gives place, with a half-life of about 90 min, to a weaker negative Cotton band, centered at about 202 mμ, as reported by Blout et al.[8] for form II. Curve b represents an intermediate stage.

The positive lobe of the initial ORD spectrum (a) is considerably weaker ($[m]_{223} = +47,500 \pm 500°$) than the negative lobe ($[m]_{208} = -90,000 \pm 500°$), both values being extrapolated back to zero time. This indicates that more than one Cotton band is involved, a conclusion which (as is often the case) is much less equivocal from the corresponding CD spectra, shown in Fig. V-4. Here, the initial spectrum (a) clearly consists of two bands: a large, positive one at 215 mμ ($\epsilon_L - \epsilon_R = 30.0$ 1-mole^{-1}-cm^{-1} at zero time) and a weaker negative band at 199 mμ ($\epsilon_L - \epsilon_R \cong -10.5$ 1-mole^{-1}-cm^{-1} at zero time). As the isomerization proceeds, the positive band becomes weaker and the initial negative band is replaced by a somewhat stronger negative band (c) at 206 mμ ($\epsilon_L - \epsilon_R = -18.0$ 1-mole^{-1}-cm^{-1}). Even after many days, a weak positive band apparently centered at about 226 mμ is still apparent. It has been reported by Carver et al.[17] that such a band is required in order to fit the observed ORD spectrum by a summation of calculated curves, although it is not apparent to casual inspection. They have directly demonstrated the existence of this band in the CD spectrum of poly-L-proline in water.

In addition to these bands, examination of the long wavelength region of the form I spectrum using a 1-mm path length (an 0.1-mm cell was normally used) reveals a weak negative band at 232 mμ, having an intensity of -0.82 1-mole^{-1}-cm^{-1} This is shown as curve a' in Fig. V-4. This has been assigned[15] as the $n - \pi^*$ band, hitherto unreported for any form of poly-L-proline. This band cannot be detected in the form II spectrum.

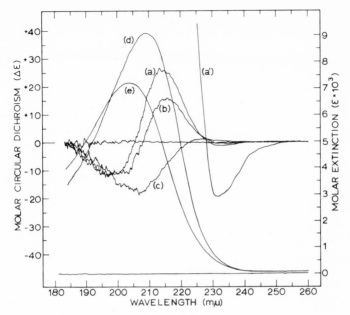

Fig. V-4. CD and UV absorption spectra of poly-L-proline (molecular weight 14,000), 1.02×10^{-2} M in trifluoroethanol for CD; 1.03×10^{-3} M for absorption measurements. Curves (*a*), (*b*), and (*c*) are run at successive times. For (*a*) and (*b*) the intervals given below represent the times after initial dissolution at which the maximum and minimum, respectively, are observed; (*a*) 13 min, 18 min; (*b*) 56 min, 60 min; (*c*) was measured at 19 hr; (*a'*) run at 10 min, using a path length of 1 mm. (*vs.* 0.1 mm for other CD curves) in order to show the $n - \pi^*$ band clearly; (*d*) ultraviolet absorption at 15 min; (*e*) ultraviolet absorption at 19 hr.

The absorption spectra (see Fig. V-4) of forms **I** and **II** are also quite distinct. In trifluoroethanol, form **I** has a maximum at 210 mμ (ϵ = 8900 at zero time), while form **II** has a maximum at 202 mμ (ϵ = 7100). These correspond to, but do not exactly coincide with, the principal bands of the CD spectrum of each form. Gratzer *et al.*[9] have observed a similar but smaller difference for aqueous solutions.

The theoretical interpretation of the optical spectra of poly-L-proline in both of its forms has been considered by Pysh.[18]

Following the general procedure of Tinoco[19] and Woody,[20] he places point monopoles on the N, C, and O atoms, and calculates dichroism as the sum of the following terms: (a) exciton splitting of the $\pi - \pi^*$ peptide transition near 200 mμ, discussed in Chapter IV; (b) a "coupled oscillator" term from simultaneous motions of electrons in two interacting groups; this involves the $n' - \pi^*$ and $n - \sigma^*$ transitions at 165 mμ and 150 mμ, respectively;[21] (c) a static term, involving all transitions still further in the ultraviolet. The absorption spectra were also calculated.

The results of these calculations are summarized in Fig. V-5. Here, the cross-hatched bars represent the absorption intensity in the split bands. It will be seen that these explain the marked blue shift as form **I** isomerizes into form **II**: intensity is shifted from the red exciton band, predicted to be polarized parallel to the helical axis, to the blue band, polarized perpendicularly to the helical axis. The sign and order of magnitude of the CD bands in form **I** (full curve) is correctly predicted; here the exciton term is strongly dominant. (The calculations do not predict that the red band will be considerably stronger than the blue, as is actually the case.) For form **II**, there is a further splitting of the blue band; this and other contributions are represented as dotted lines. The full line is the net ellipticity. It reproduces the form of the experimental spectrum quite well.

Pysh also calculated the intensities (but not the positions) of the $n - \pi^*$ CD contributions. For form **I**, this was predicted to be two orders of magnitude weaker than the $\pi - \pi^*$ bands; it is actually more nearly one order of magnitude weaker. For form **II**, he correctly predicted a still weaker intensity for the $n - \pi^*$ band.

These calculations, based on the X-ray structures, are qualitatively and even semiquantitatively in agreement with experiment, and may be taken as support for the conclusion that the solution conformations are very similar to those in the solid state.

The mechanism of the **I** $\rightleftharpoons$ **II** transformation in poly-L-proline has been discussed by several authors.[10,22,23] There

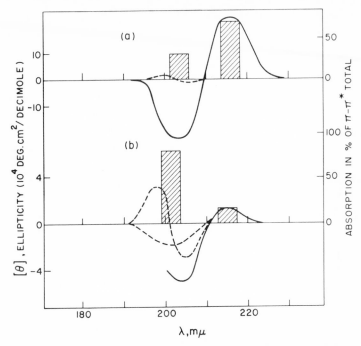

Fig. V-5. Schematic representation of the calculated absorption and circular dichroism spectra of poly-L-proline I and II, according to Pysh (ref. 18). The bars represent oscillator strength, given as percent of the total $\pi - \pi^*$ oscillator strength. The dotted lines show separate contributions to rotational strength, and the solid lines are the approximate net ellipticity curves for a band width of *ca.* 2500 cm^{-1}.

seems to be little doubt that the transition is reversible and highly cooperative in nature. There is no evidence at present for any intermediate form. The observed activation energy of about 20 kcal[24,25] undoubtedly corresponds to the rotational barrier about the C—N bond, which we have already discussed. Engel[22] has observed the transition in *n*-butanol-benzyl alcohol mixtures (these solvents support form **I** and form **II** respectively) and has shown by equilibrium measurements that the transition becomes more abrupt with increasing molecular weight, in a manner similar to the α-helix-random coil transition. Such behavior is characteristic of a reaction in which many monomer units must act in concert.

2. *Poly-L-Acetoxyproline.* Poly-3-hydroxy-L-proline and its acetyl derivative, poly-L-acetoxyproline

have two forms in the crystalline state, the conformations of which appear to be very similar to those of the corresponding forms of poly-L-proline,[26,27] and undergo similar mutarotations in appropriate solvents. In Fig. V-6 are shown the CD spectra of poly-L-acetoxyproline (DP *ca.* 100) in trifluoroethanol solution at 30°, reflecting the transformation of form **I** to form **II**. The reaction is somewhat slower than for poly-L-proline, the half-life of form **I** being *ca.* 120 min. The CD spectra of form **I** (*a*) and form **II** (*b*) are in general similar to those of the corresponding forms of poly-L-proline, but differ in that the negative band at 230 mμ, assigned to the $n - \pi^*$ transition of form **I**, is three times more intense in poly-L-acetoxyproline. In form **II**, the negative $\pi - \pi^*$ band is considerably weaker and the positive band (at 226 mμ) considerably stronger than for form **II** of poly-L-proline. The positive band of form **II** and the $n - \pi^*$ band of form **I** are shown more clearly in Fig. V-7, in which a 1-mm path length was used, as in spectrum *a'* of Fig. V-5.

The absorption spectra of form **I** [(*d*) in Fig. V-6] and form **II** [(*e*) in Fig. V-6] resemble very closely the corresponding spectra for poly-L-proline. The ester group apparently makes a negligible contribution in this region of the spectrum.

Fasman[28] has reported the ORD spectrum of a film of poly-L-acetoxyproline; the results appear to be qualitatively in accord with our CD measurements on the solution. In the ORD spectra of poly-L-acetoxyproline in dichloromethane solution (a solvent which absorbs strongly below about 230

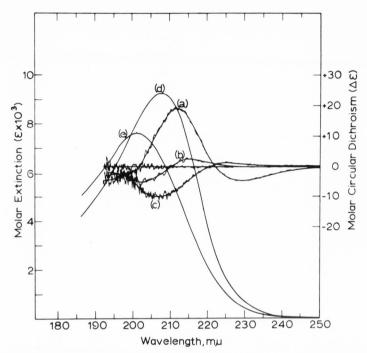

Fig. V-6. CD and ultraviolet absorption spectra of poly-L-acetoxyproline (mol. wt. 11,500), 1.00×10^{-2} M in trifluoroethanol for CD, and 8.00×10^{-4} M in trifluoroethanol for absorption measurements. Curves (a), (b), and (c) are run at successive times at 30°C, using a path length of 0.1 mm. For curves (a) and (b), the intervals given below represent the times after initial dissolution at which the two major extrema, proceeding from long to short wavelength respectively, are observed: (a) 24 min, 27 min; (b) 182 min, 186 min; (c) 26 hr. The absorption curves (d) and (e) were run at 12 min and 26 hr respectively, using a 1-mm path length.

mμ, making measurements impossible at shorter wavelengths) he reported a weak positive Cotton band centered at 254 mμ for form **I** and a weak negative Cotton band centered at 233 mμ for form **II**. We have not observed corresponding bands in trifluoroethanol solution. A band as far to the red as 254 mμ is difficult to explain for this polymer.

    3. *Poly-N-Methyl-L-Alanine.* Goodman *et al.*[29-31] have

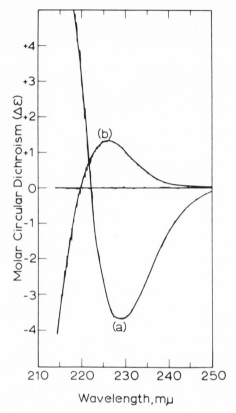

Fig. V-7. CD spectra of poly-L-acetoxyproline, 1.00×10^{-2} M in trifluoro-ethanol. Curves (a) and (b) run at 12 min and 26 hr after dissolution, respectively, were recorded, using a path length of 1 mm in order to more clearly show the weaker long-wavelength extrema.

prepared poly-N-methyl-L-alanine

$$\begin{array}{ccc} CH_3 & CH_3 & O \\ | & | & || \\ \end{array}$$
$$\left[N \!-\!\!-\!\! CH \!-\!\! C \right]_n$$

from the N-carboxy anhydride and have carefully studied it by both NMR and ORD—CD. They have also studied the model compound, N-acetyl-N-methyl-L-alanine methyl ester:

$$
\begin{array}{ccc}
& CH_3 & CH_3 \\
& | & | \\
CH_3CON & \!\!\!\!\!\!\!\!\! & \!\!\! CHCO_2CH_3
\end{array}
$$

This polypeptide is similar to poly-L-proline but may be expected to have greater conformational freedom. It can be thought of as poly-L-proline with the five-membered ring cut by removal of the central methylene group, thereby permitting at least some rotational freedom about the N—C bond (called the φ bond; see later discussion).

In Fig. V-8 are shown the NMR spectra of the polymer (a)

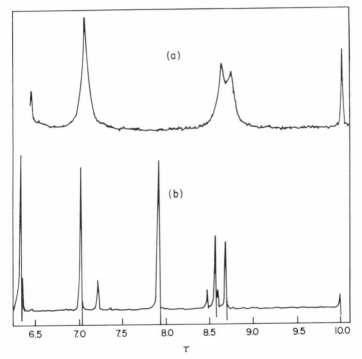

Fig. V-8.   60 MHz NMR spectra of (a) poly-N-methyl-L-alanine and (b) N-acetyl-N-methyl-L-alanine methyl ester in trifluoroethanol solution.

and the model compound (*b*). In the latter, the N-methyl resonance appears as a large peak at 7.0τ and a smaller peak at 7.2τ; the C-methyl group appears as two doublets near 8.5τ. The reason for the appearance of two sets of peaks for each methyl group is that, like a prolyl unit, this compound exists in both *cis* and *trans* conformations:

These are defined in the sense appropriate to the polymer chain. The exchange between them is slow on the NMR time scale, the barrier being probably about 20 kcal. From the findings of Anet and Bourn,[32] who showed by means of the nuclear Overhauser effect that in *N*,*N*-dimethylformamide the most shielded methyl group is that *cis* to the carbonyl group, the strongest peaks are assigned to the *trans* conformer; it thus predominates to an extent of about 4:1.

In the spectrum of the polymer (*a*) only one N-methyl and one C-methyl resonance are observed, and these are in the positions corresponding to the *trans* conformation. Thus, unlike the model compound, the polymer is all *trans*. This behavior was found to persist in dichloromethane and a number of other solvents. (In trifluoroacetic acid, however, both *cis* and *trans* resonances were observed; this is discussed further in the next section.)

This finding suggests that the polymer very probably has a fixed, helical conformation. This was further supported by a marked hypochromism ($\epsilon = 5600$ at 201 mμ), characteristic of strongly interacting chromophores in a helix. The CD spectrum of the polymer, (*a*) in Fig. V-9, had a very broad, negative band centered at 223 mμ, followed by a narrower positive band at 192 mμ. This was interpreted as being very

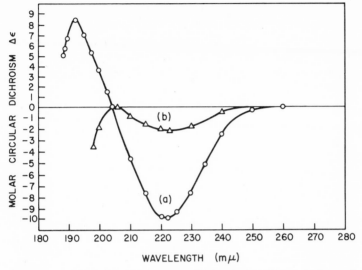

Fig. V-9. Circular dichroism spectra of (a) poly-N-methyl-L-alanine and (b) N-acetyl-N-methyl-L-alanine methyl ester in trifluoroethanol solution.

similar to the CD spectrum of the α-helix (Fig. IV-7), i.e., a negative $n - \pi^*$ and $\pi - \pi^*$ band followed by a positive $\pi - \pi^*$ band. The first two bands are so close together that two separate minima cannot be distinguished.

The spectrum of the model compound shows only a weak $n - \pi^*$ band, followed by the beginning of a negative $\pi - \pi^*$ band. This gives further support for a helical polymer conformation, for if the polymer were a random coil, such a marked contrast in the CD spectra of the model and the polymer would not be expected.

The present state of the theory of the optical spectra of polypeptide chains does not enable one to deduce structure from experimental observations. The NMR spectrum provides a valuable clue, but gives no detailed conformational information. A further recourse is to calculate the potential energies of all possible helical conformations, and see which ones have the lowest. This has now been done by several authors for a number of polypeptides, including poly-L-proline.[13] Mark and Goodman[3031] have extended these calculations to poly-N-methyl-L-alanine. In considering the rotational states of a polypeptide chain, one must consider three main-chain bonds, which by convention[33] are designated as shown below:

The ω-bond is fixed, the $-\overset{\overset{\textstyle O}{\|}}{C}-N\diagdown$ group being assumed to be planar. For α-helices it is *trans* (defined as 0°); for poly-N-methyl-L-alanine, Mark and Goodman assumed a *trans* conformation on the basis of the NMR results. With this dihedral angle fixed, one may then express any and all helical conformations by specifying the dihedral angles of the φ and ψ bonds. For a planar zigzag, these are taken as zero. Other angles are defined by, as it were, standing on the α-carbon atom and looking toward the nitrogen; as one rotates the planar peptide unit clockwise away from the planar zigzag state, φ increases from zero. One then looks toward the carbonyl carbon (while still standing on the α-carbon) and goes through increasing positive values of ψ by rotating this peptide unit clockwise away from the planar zigzag state. In these terms, the right-handed α-helix is specified by $\varphi = 132°$ and $\psi = 123°$; a left-handed α-helix by the complements of these angles; $\varphi = 228°$ and $\psi = 237°$. Conformational energies may be expressed as a contour map with φ and ψ as the ordinates in the horizontal plane.

One must then choose suitable functions for the torsional potentials of the φ and ψ bonds and for the van der Waals interactions between pairs of nonbonded atoms. For the φ and ψ bonds, Mark and Goodman assumed potential minima at 0° and ±120° with barrier heights of 1.5 and 1.0 kcal per mole, respectively, at $\pm 60°$ and $180°$.[34] For the van der Waals interactions, they used standard bond angles and lengths and a Lennard-Jones "6-12" potential function:

$$V_{ij} = d_{ij}/r_{ij}^{12} - e_{ij}/r_{ij}^6$$

The values of d and e were those employed by Scott and Scheraga.[35] The potential-energy contour map was generated by employing a computer program to calculate the potential energies for values of φ and ψ from $0°$ to $360°$, varying each in $10°$ increments; a total of about 1200 points being thus obtained. Points of equal energy were then connected by contour lines, drawn in 1-kcal increments. Points representing energies greater than 5 kcal were ignored. The "zero" of energy is actually not zero, but is instead arbitrarily taken as the fifth contour below the 5-kcal level.

The result of these calculations is shown in Fig. V-10. The striking degree of restriction introduced by the presence of two methyl groups is evident in the fact that regions representing 5 kcal/mole or less in energy take up only 2.5% of the available conformational space. Right-handed α-helical conformations are entirely ruled out by methyl-methyl contacts. The four discernible energy minima are described in Table V-1; here n is the number of residues per turn and d is the residue translation.

TABLE V-1.   **Low-Energy Conformations of the Poly-N-methyl-L-Alanine Chain**

Conformation	φ	ψ	E, kcal/mole	n	d, Å[a]
I	$30°$	$250°$	-0.9	3.24	2.48
II	$210°$	$250°$	-0.3	3.59	-1.67
III	$80°$	$345°$	2.5	2.91	-3.35
IV	$240°$	$345°$	1.5	4.77	2.51

[a]Negative values of d indicate left-handed helices.

Conformation **I**, apparently the most probable one, is an approximately 3-fold right-handed helix. Stereo views of this structure are shown in Fig. V-11. Conformation **II** is similar

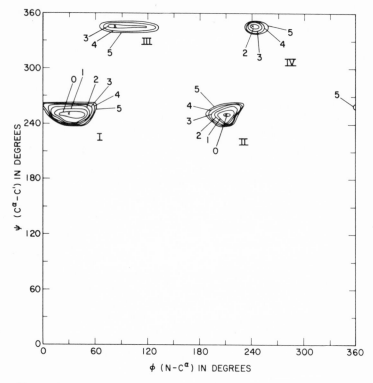

Fig. V-10. Potential-energy contours for helical conformations of poly-*N*-methyl-L-alanine.

to a left-handed α-helix. Conformation **III** is a left-handed helix similar to that of poly-L-proline **II**. Conformation **IV** is a right-handed, approximately five-fold helix. These latter two conformations are unlikely to resemble the true one. Consideration of peptide dipole-dipole interactions indicates that **IV** is probably of even higher energy, while **I, II,** and **III** would be further stabilized. A clear choice between **I** and **II** does not seem possible from energy calculations. The final decision will, of course, be supplied by X-ray diffraction.

Similar energy calculations have also been reported by Liquori and De Santis.[36]

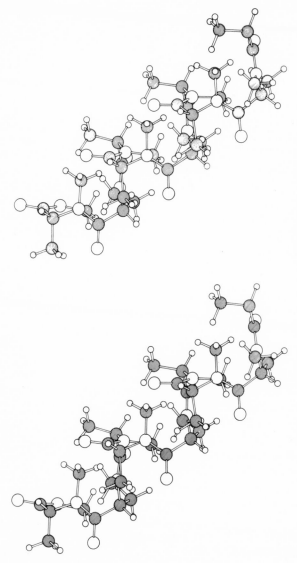

Fig. V-11. Stereo views of the type **I** helix of poly-L-methyl-L-alanine.

4. *Polysarcosine.* Comparatively little attention has been given to polysarcosine (poly-*N*-methylglycine):

$$\left[-N-CH_2-\overset{\displaystyle O}{\overset{\displaystyle \|}{C}}- \atop \underset{\displaystyle CH_3}{|} \right]_n$$

Probably one of the principal reasons for this neglect is that polysarcosine (which is isomeric with polyalanine) has no asymmetric carbon atoms and hence is not subject to optical rotatory dispersion or circular dichroism studies. The sarcosine unit is present in certain proteins and natural polypeptides but its occurrence is rather rare.

Fessler and Ogston[37] measured the viscosity, sedimentation, and diffusion of polysarcosine in aqueous solution and concluded that it was probably a random coil. Glazer and Rosenheck[38] reached a similar conclusion from a study of its ultraviolet spectrum, and believed that the absence of hydrogen bonding was responsible. We have seen, however, that this is not an adequate criterion, since poly-L-proline and poly-*N*-methyl-L-alanine assume helical conformations despite a lack of stabilizing hydrogen bonds. It was therefore of some interest to investigate further the conformation of polysarcosine, which is the simplest polypeptide of this type and in which steric restrictions might be expected to be the least demanding.

In Fig. V-12 are shown the 60 MHz spectra of polysarcosine of DP 94 (*a*) and of the model compound *N*-acetylsarcosine methyl ester (*b*)[39]

$$CH_3CONCH_2CO_2CH_3 \atop \underset{\displaystyle CH_3}{|}$$

employed to assist in making assignments in the polymer spectrum; both spectra were observed in d_6-DMSO at *ca.* 35°, using 2% tetramethylsilane as internal reference. (The multi-

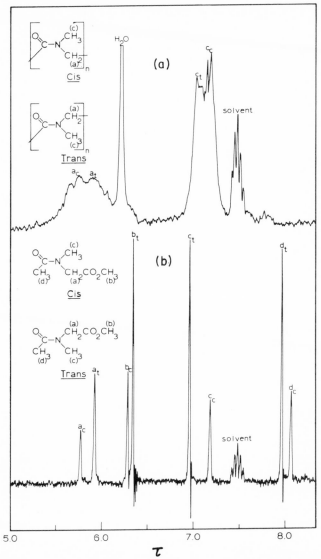

Fig. V-12.   60 MHz NMR spectra of (*a*) polysarcosine (DP 94) and (*b*) *N*-acetylsarcosine methyl ester in d_6-DMSO (*ca.* 10% w/v) at 35°. The peak assignments are indicated by letters, the subscripts denoting the *cis* and *trans* conformers.

plet at 7.47τ is due to residual d_5-DMSO.) In the model compound spectrum, it is observed that all the peaks are doubled, giving a total of eight, although there would at first sight appear to be only four different kinds of protons. The reason for this of course is that, like the corresponding model compound for poly-N-methyl-L-alanine, the compound can exist in both *cis* and *trans* conformations, which exchange slowly on the NMR time scale:

The peak assignments are given in Fig. V-12. It can be seen from the relative intensities of all peaks in (*b*) that the *trans* conformer is preferred by about 2.5 to 1 in d_6-DMSO. (Relative *cis* and *trans* peak heights vary slightly for the a, b, c and d protons because of differential broadening by weak, unresolved coupling.) The *trans*/*cis* ratio varies somewhat in other solvents (d_4-methanol, CDCl₃, trifluoroacetic acid, trifluoroethanol, and pyridine) but the *trans* is always strongly preferred.

Turning to the polymer spectrum (*a*), we find the N-methyl and methylene peaks seemingly much broadened. [There are of course no peaks corresponding to the b and d protons of spectrum (*b*).] The N-methyl region shows clear evidence of more than two peaks; the methylene region also shows this higher multiplicity, but is more difficult to interpret. It becomes better resolved in the 220 MHz spectrum (Fig. V-13).

In Fig. V-14, the N-methyl multiplet of Fig. V-13 is expanded 5-fold. One can now distinguish seven peaks by inspection; analysis of the multiplet with a duPont model 310 Curve Analyzer makes it clear that there must be at least one additional peak to match the observed spectrum. A set of peak positions and relative intensities (*i.e.,* areas) which matches the observed spectrum within the probable experimental error is indicated as "sticks" beneath the spectrum.

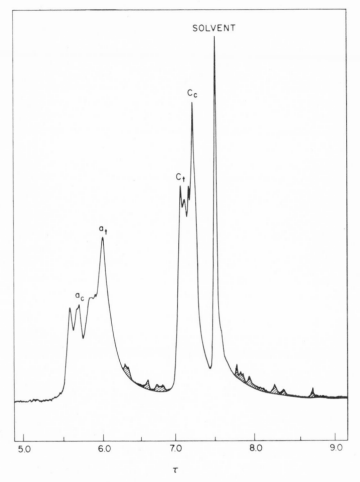

Fig. V-13.   220 MHz spectrum of polysarcosine (DP 94) in d_6-DMSO at
35°. (The small cross-hatched peaks are spinning sidebands.)

It is believed that these peaks reflect the fact that N—CH_3
shielding is not only a function of the ω-bond conformation
of the peptide unit to which it is attached but also of the conformations of the nearest neighbors as well.   We can then

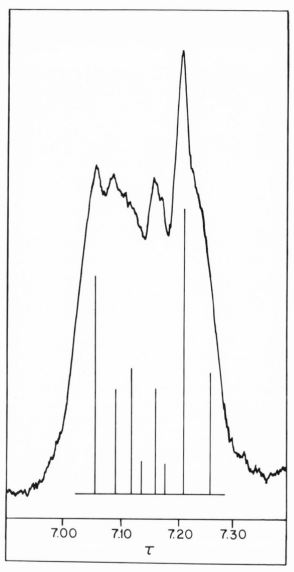

Fig. V-14.   Expanded 220 MHz N-methyl spectrum of polysarcosine under the same conditions as for Fig. V-13. The "stick" spectrum is the result of an analysis of the experimental band intensities (see text).

recognize eight species of peptide triad sequences:

trans-trans-trans	*cis-cis-cis*
trans-trans-cis	*cis-cis-trans*
cis-trans-trans	*trans-cis-cis*
cis-trans-cis	*trans-cis-trans*

If the chain were truly random with respect to this bond, the spectrum would consist of eight equal peaks. This clearly is not the case. Yet if we assume that the four upfield peaks correspond to sequences with *cis* central units and the four downfield peaks to sequences with *trans* central units, it is found that the probabilities of *cis* and *trans* units are nearly equal, the *trans/cis* ratio being *ca.* 0.9. This contrasts to the strong *trans* preference of the model compound.

It is difficult to make specific triad peak assignments. It appears that certain triad sequences tend to be favored and others excluded, the energy differences being as great as *ca.* 1 kcal in some cases. Thus, not surprisingly, the ω-bond conformation of a peptide unit influences that of its neighbors to some degree. Molecular models do not suggest severe steric limitations for any of these sequences, and so it will be of interest to see what potential-energy calculations predict. These have not yet been carried out. One might conjecture that the *cis-cis-cis* and *trans-trans-trans* conformations would have the lowest potential energy, since the latter appears to be exclusively preferred for poly-*N*-methyl-L-alanine (Sec. 3) and poly-L-proline (Sec. 1) can exist in both conformations, depending upon the solvent.

The ω-bond conformational preference of polysarcosine varies markedly with solvent. (All measurements were made at 35°.) In d_4-methanol the spectrum is similar to that in d_6-DMSO. In $CDCl_3$, the *cis* conformation is strongly preferred. In pyridine, the *trans* conformation is of somewhat lower energy. In trifluoroacetic acid, trifluoroethanol, and water the *trans* conformation is preferred. However, in all these solvents, fine structure due to neighboring peptide conformations is either more difficult to discern than in d_6-DMSO, or is entirely absent.

Recently, Goodman and Prince[40] have made analogous observations of multiple methyl peaks for poly-N-methyl-L-alanine in trifluoroacetic acid. It thus appears that this polymer, which (as we have seen in Sec. 3) probably exists in a single helical conformation in solvents such as trifluoroethanol and methylene chloride, has a random-coil conformation in trifluoroacetic acid, and thus behaves in a manner analogous to poly-α-amino acids which are not alkylated at the amino nitrogen.

REFERENCES FOR CHAPTER V

(1) A. Allerhand and H. S. Gutowsky, *J. Chem. Phys.,* **41,** 2115 (1964).

(2) P. M. Cowan and S. McGavin, *Nature,* **176,** 501 (1955).

(3) V. Sasisekharan, *Acta Cryst.,* **12,** 897 (1959).

(4) W. Traub and V. Shmueli, *Nature,* **198,** 1165 (1963).

(5) J. Kurtz, A. Berger, and E. Katchalski, *Nature,* **178,** 1066 (1956).

(6) W. F. Harrington and M. Sela, *Biochim. Biophys. Acta,* **27,** 24 (1958).

(7) I. Z. Steinberg, A. Berger, and E. Katchalski, *ibid.,* **28,** 647 (1959).

(8) E. R. Blout, J. P. Carver, and J. Gross, *J. Am. Chem. Soc.,* **85,** 644 (1963).

(9) W. B. Gratzer, W. Rhodes, and G. D. Fasman, *Biopolymers,* **1,** 319 (1963).

(10) F. Gornick, L. Mandelkern, A. F. Diorio, and D. E. Roberts, *J. Am. Chem. Soc.,* **86,** 2549 (1964).

(11) W. F. Harrington and P. von Hippel, *Advances in Protein Chemistry,* **16,** 1 (1961).

(12) J. P. Carver and E. R. Blout, in "Treatise on Collagen," ed by G. N. Ramachandran, Academic Press, London, Vol. I, 1967.

(13) P. R. Schimmel and P. J. Flory, *Proc. Nat. Acad. Sci. U.S.,* **58,** 52 (1967).

(14) F. A. Bovey and F. P. Hood, *J. Am. Chem. Soc.,* **88,** 2326 (1966).

(15) F. A. Bovey and F. P. Hood, *Biopolymers,* **5,** 325 (1967).

(16) E. R. Blout and E. Schechter, *Biopolymers,* **1,** 565 (1963).

(17) J. P. Carver, E. Schechter, and E. R. Blout, *J. Am. Chem. Soc.,* **88,** 2550 (1966).

(18) E. S. Pysh, *J. Mol. Biol.,* **23,** 587 (1967).

(19) I. Tinoco, *Advances in Chemical Physics,* **4,** 113 (1962).

(20) R. W. Woody, Ph.D. Thesis, University of California, Berkeley (1962).

(21) D. L. Peterson and W. T. Simpson, *J. Am. Chem. Soc.,* **79,** 2375 (1957).

(22) J. Engel, *Biopolymers,* **4,** 945 (1966).

(23) J. Applequist, *Biopolymers,* **6,** 117 (1968).

(24) I. Z. Steinberg, W. F. Harrington, A. Berger, M. Sela, and E. Katchalski, *J. Am. Chem. Soc.,* **82,** 5263 (1960).

(25) A. R. Downie and A. A. Randall, *Trans. Faraday Soc.*, **55**, 2132 (1959).

(26) V. Sasisekharan, *J. Polymer Sci.*, **47**, 391 (1960).

(27) V. Sasisekharan, *Acta Cryst.*, **12**, 903 (1959).

(28) G. D. Fasman, *Biopolymers*, **4**, 509 (1966).

(29) M. Goodman and M. Fried, *J. Am. Chem. Soc.*, **89**, 1264 (1967).

(30) J. E. Mark and M. Goodman, *J. Am. Chem. Soc.*, **89**, 126 (1967).

(31) J. E. Mark and M. Goodman, *Biopolymers*, **5**, 809 (1967).

(32) F. A. L. Anet and A. J. R. Bourn, *J. Am. Chem. Soc.*, **87**, 5250 (1965).

(33) J. T. Edsall, P. J. Flory, J. C. Kendrew, A. M. Liquori, G. Nemethy, G. N. Ramachandran, and H. A. Scheraga, *Biopolymers*, **4**, 121 (1966).

(34) D. A. Brant, W. G. Miller, and P. J. Flory, *J. Mol. Biol.*, **23**, 47 (1967).

(35) R. A. Scott and H. A. Scheraga, *J. Chem. Phys.*, **45**, 2091 (1966).

(36) A. M. Liquori and P. De Santis, *Biopolymers*, **5**, 815 (1967).

(37) J. H. Fessler and A. G. Ogston, *Trans. Faraday Soc.*, **47**, 667 (1951).

(38) A. N. Glazer and K. Rosenheck, *J. Biol. Chem.*, **237**, 3674 (1962).

(39) F. A. Bovey, J. J. Ryan, and F. P. Hood, *Macromolecules*, **1**, 305 (1968).

(40) M. Goodman and F. Prince, Jr., *Macromolecules*, in press.

Author Index

Subject Index